Österreichische Akademie der Wissenschaften
Schriftenreihe der Erdwissenschaftlichen Kommission
Band 1

Metallogenetische und Geochemische Provinzen
Metallogenetic and Geochemical Provinces

Symposium Leoben, November 1972

Herausgegeben von / Edited by

W. E. Petrascheck

Wirkliches Mitglied der Österreichischen Akademie der Wissenschaften
Vorstand des Instituts für Geologie und Lagerstättenlehre
der Montanistischen Hochschule Leoben

Springer-Verlag Wien GmbH

Symposium, organisiert im Hinblick auf
das Internationale Geologische Korrelationsprogramm (IGCP)
Symposium organized with regard to
the International Geological Correlation Program (IGCP)

Veröffentlicht unter Mitwirkung und
mit finanzieller Unterstützung der UNESCO
Published in collaboration with
and with the financial support of UNESCO

Mit 48 Abbildungen und 17 Tabellen

ISBN 978-3-211-81249-5   ISBN 978-3-7091-4065-9 (eBook)
DOI 10.1007/978-3-7091-4065-9

Inhaltsverzeichnis

Vorwort

Das Motiv für die Veranstaltung dieses Symposiums über die Beziehungen zwischen metallogenetischen und geochemischen Provinzen war ein egoistisches Informationsbedürfnis des Organisators. Als ich 1967 an dem Manuskript über Erzprovinzen und Kontinentaldrift arbeitete, ergab sich die Frage, ob die Metalle der Lagerstätten in den wandernden Kontinentalschollen beheimatet sind oder ob sie aus dem wechselnden Substratum des oberen Mantels und der daraus gespeisten ozeanischen Kruste stammen. Es galt also, das Material zu überprüfen, das die Beziehungen der Erzlagerstätten zur Geochemie ihrer kontinentalen Umgebung oder zur Geochemie des Ozeanbodens zeigt. Dies kann zugleich auch Beiträge ergeben zu dem Problem der modernen lateralsekretionären Lagerstättenbildung.

Die wirkungsvollste Art, sich bei der heutigen großen Menge von Literatur über ein komplexes Thema zu informieren, ist, kompetente Leute zusammenzubitten, damit sie darüber vortragen und diskutieren. Ich darf sagen, daß meinem Bedürfnis voll Genüge geleistet wurde und ich habe dafür all den lieben Kollegen, die zum Teil von sehr weit nach Leoben gekommen sind, herzlichen Dank zu sagen. Alle Vorträge behandeln, ja man kann fast sagen, klären in ihrer Gesamtheit das aufgeworfene Problem.

Ich möchte aber doch glauben und hoffen, daß nicht nur der Veranstalter, sondern auch die Teilnehmer von den Vorträgen und Diskussionen Nutzen hatten. Ein intensiver Gedankenaustausch war möglich, da die Teilnehmerzahl mit Absicht klein gehalten und auf spezielle Sachkenner beschränkt war und da das Programm im Gegensatz zu vielen Kongressen weite Lücken für Diskussionen ließ.

Zwei Exkursionen zu typischen Lagerstätten Österreichs, der Bleizinklagerstätte Bleiberg und der Sideritlagerstätte Erzberg, die — wie könnte es anders sein — nun auch zwischen Hydrothermalisten und Syngenetisten hin- und hergezerrt werden, waren für unsere ausländischen Gäste bestimmt und durch sonniges Herbstwetter begünstigt zu dieser späten Jahreszeit. Den gastlichen Bergbaugesellschaften und ihren ausgezeichneten geologischen Führern galt der besondere Dank der Besucher.

Die Vorbereitung der Tagung erforderte einige Zeit, die seit der ersten informativen Korrespondenz mit den dazu gebetenen Kollegen bis zur Durchführung im November 1972 4 Jahre dauerte. Dabei bot natürlich der „finanzielle background" ein größeres Problem als der geochemische background der Lagerstätten. Das Zustandekommen ist durch großzügige Zuwendungen durch die UNESCO und des österreichischen Bundesministeriums für Wissenschaft und Forschung, durch die finanzielle Mitwirkung der International Association of Geochemistry and Cosmochemistry (Professor Earl Ingerson), durch die Beistellung der Einrichtungen der Montanistischen Hochschule Leoben und durch die Gastfreundschaft der Bleiberger Bergwerks Union und der Österreichischen Alpine-Montan-Gesellschaft ermöglicht worden. Herr Bürgermeister Leopold Posch gab im Namen der Stadtgemeinde Leoben ein festliches Abendessen.

Die Abstracts der Tagung erschienen relativ bald nachher in den Mineralium Deposita (Vol. 8, Nr. 1/1973) dank dem Entgegenkommen von Herrn Professor Maucher; der vorliegende Band wurde durch den hohen Druckkosten-Beitrag der österreichischen Akademie der Wissenschaften in Wien, die eine eigene Kommission für das IGCP gegründet hat, in dankenswerter Weise ermöglicht. Herrn Dozent Dr. W. Frisch (Leoben) danke ich sehr für wertvolle Redaktionsarbeit.

Es ist zu hoffen, daß die aus dem Symposium erwachsenen Anregungen zu einer systematischen regionalen und globalen Korrelation der geochemischen und metallogenetischen Daten führen. Es soll dies zugleich auch ein erster Beitrag Österreichs zum Internationalen Geologischen Correlations-Programm sein.

Leoben, Juli 1973

W. E. Petrascheck

Preface

The motivation for organizing a symposium on the relations between metallogenic and geochemical provinces was just an egoistic desire of the organizer for information. In 1967, when I was writing the manuscript for the article: „Ore Provinces and Continental Drift", the question arose whether the ore metals of the deposits originate in the wandering parts of the continental crust or in different parts of the subjacent upper mantle or the oceanic crust, respectively. It became necessary to examine the relation between the mineral deposits and the geochemical background of their continental environment or of the ocean floor. This can contribute also to the problem of the modern lateral-secretionary metallogenesis.

Abb. 1. Der Bürgermeister von Leoben, Direktor L. *Posch* mit
Prof. W. E. *Petrascheck* bei der Eröffnung des Banketts.

Abb. 2. Festbankett, gegeben von der Stadt Leoben.

Regarding the tremendous amount of literature on a complex subject, the best way to get information is asking a number of competent people to come together for reporting an discussing about this subject. I can say that my desire was satisfied completely, and I have to express my sincere thanks to all those dear collegues who came, partly from very far, to Leoben. One can even say that in their entirety all the lectures referred did almost clarify the problem.

But I would hope that not the organizer alone but also the participants could benefit from the papers and the discussions. An intense exchange of opinions was possible because the number of the participants was comparatively small and limited purposely to people specialized in the matter, and because the time schedule of the program let ample gaps for discussion.

Two excursions to typical ore deposits of Austria, the lead-zinc mine of Bleiberg and the siderite mine of the Erzberg which — how could it be else at present time — are now teared between hydrothermalists and syngenetists, have been organized for our foreign guests. They were favoured by a brilliant sunny weather at this late autumn season. The visitors were particularly obliged to the excellent geological guides and for the hospitality of the mining companies.

The preparation of the meeting has needed considerable time from the first informative correspondence with the colleagues asked to participate until the opening of the session. And, of course, the „financial background" of the symposium has been a greater problem than the geochemical background of ore formation.

The realization of the meeting was made possible by generous support of UNESCO, of the Austrian Bundesministerium für Wissenschaft und Forschung, by financial assistance of the International Association of Geochemistry and Cosmochemistry (Professor Earl Ingerson), by the disposition of the Institutions of the Montanistische Hochschule, and the hospitality of the Bleiberger Bergwerks-Union and the Österreichische Alpine Montangesellschaft. The mayor of Leoben, Mr. Leopold Posch, gave a splendid dinner to the participants.

The abstracts of the symposium have been published rather soon after the meeting in Mineralium Deposita (Vol. 8, Nr. 1, 1973) due to the kindness of the editor of this periodical, Professor A. Maucher. The printing of the present volume has been enabled by a high contribution of the Austrian Academy of Sciences in Vienna which has established a special commission for the International Geological Correlation Program. I have to thank Dozent Dr. W. Frisch for valuable edition work.

It is hoped that the suggestions accrued from this symposium might lead to a systematic collection and correlation of metallogenic and geochemical data in a regional as well as in a world-wide scale. This shall be at the same time a first contribution of Austria to the International Geological Correlation Program.

Leoben, July 1973

W. E. Petrascheck

Distribution of Metallogenic Provinces in Relation to Major Earth Features*

Philip W. *Guild***

Zusammenfassung

„Die Verteilung metallogenetischer Provinzen in Bezug auf Großeinheiten der Erde"

Aus der Theorie und den Beobachtungen in Zusammenhang mit der globalen Tektonik ergeben sich neue Wege 1) für die Abgrenzung und Verteilung einiger metallogenetischer Provinzen bezogen auf die Platten in ihrer heutigen Position; 2) bei der Aufstellung von möglichen, unterschiedlichen genetischen Modellen; und 3) bei der Analyse der Verteilung älterer Lagerstätten in Bezug auf a) ihre ursprüngliche Zusammengehörigkeit oder b) die Lage und Beschaffenheit früherer Plattenränder.

Es gibt im wesentlichen zwei Möglichkeiten: 1) Lagerstätten wurden an einem Plattenrand oder in der Nähe desselben oder 2) innerhalb von Platten gebildet. Jede der beiden Möglichkeiten läßt sich weiter unterteilen: 1 a) an konstruktiven (wachsenden), 1 b) konservativen (transformativen) oder 1 c) destruktiven (verschluckenden) Rändern; 2 a) innerhalb ozeanischer Plattenteile, 2 b) an wandernden Kontinentalrändern oder 2 c) innerhalb kontinentaler Plattenteile. Ziemlich klare Beispiele für 1 c sind die Andine (Kontinent/Ozean) und die Westpazifische (Ozean/Ozean oder Inselbogen) Provinz. Zweifelsfreie Beispiele für 1 a sind relativ selten, für 1 b völlig unsicher. Bekannte endogene Lagerstätten der Kategorie 2 sind naturgemäß auf Kontinentalgebiete beschränkt (2 c); es ist hingegen unwahrscheinlich, daß es überhaupt welche für 2 a oder 2 b gibt.

Diese Einteilung ist zu sehr vereinfacht, um allen Möglichkeiten Rechnung tragen zu können. So findet man z. B. in einem großen Gebiet der Cordillieren in den westlichen Vereinigten Staaten Merkmale von 1 c und 2 c vereint; die Unter-

---

* Publication authorized by the Director, U.S. Geological Survey.
** Prof. Dr. Ph. W. *Guild*, U.S. Geological Survey, Washington, D.C. 20244, USA.

strömung hörte ohne Zweifel zu dem Zeitpunkt auf, sobald der Mittelpazifische Rücken überfahren war, doch wurden die Verteilung und der Charakter älterer Lagerstätten ebenso von Strukturen der reaktivierten Plattform beeinflußt. Daraus läßt sich zwingend schließen, daß die mafitischen und alkalisch-mafitischen Lagerstätten von 2 c tatsächlich verhinderten wachsenden Plattenrändern (1 a) in kontinentaler Umgebung zugehören. Es ergeben sich ohne Zweifel noch weitere Komplikationen.

Die hier wiedergegebene Einteilung legt nur noch einmal in neuen Worten lange bekannte Begriffe dar, ohne ein geeignetes Mittel darzustellen, um die Ergebnisse aus der metallogenetischen Forschung mit neuen Informationen und Theorien aus anderen Gebieten der Erdwissenschaften zu korrelieren.

Abstract

Global tectonic fact and theory provide new ways of 1) categorizing the known distribution of some metallogenetic provinces with respect to the present lithospheric plates; 2) restricting possible alternative genetic models; and 3) analyzing distributional patterns of older deposits with respect to a) original contiguity, or b) the positions and nature of former plate boundaries.

In the simplest terms, deposits may form at or near 1a) accreting, 1b) transform, or 1c) consuming margins of plates; or 2a) within oceanic parts, 2b) at trailing continental margins, or 2c) within continental parts of plates. Some deposit types have clearly definable positions; others combine feature of two positions, either sequentially or simultaneously.

Analysis of mineral-deposit data in plate-tectonic terms may give new insight into such problems as mantle vs crustal provenance of ore elements and the energy to concentrate them.

Introduction

Global tectonic fact and theory are stimulating revisions in „traditional" ways of looking at geology. Metallogeny, „the study of the genesis of mineral deposits, with emphasis on their relationship in space and time to regional petrographic and tectonic features", (AGI, 1972, p. 445) is no exception (*Guild,* 1971, 1972a, 1972b, 1973; *Mitchell* & *Garson,* 1972; *Sawkins,* 1972; *Sillitoe,* 1972a, 1972b, 1972c). The restriction of many types of geologically young deposits to regions of subduction *(Guild,* 1972a) has led various workers to construct models for the genesis of porphyry copper *(Mitchell* & *Garson,* 1972; *Sawkins,* 1972; *Sillitoe,* 1972b) and other sulfide deposits. If the global tectonics is in fact a valid concept as the mass of rapidly accumulating evidence seems to suggest, most if not all ore deposits will have characteristic relationships to the lithospheric plates that can be useful: (1) to categorize them; (2) perhaps to restrict alternative genetic models; and (3) to analyze the distributional patterns of older deposits with respect to (a) original contiguity *(Petrascheck,* 1968; *Schuiling,* 1967), or (b) the positions and nature of former plate boundaries *(Guild,* 1973). The emphasis here will be on the first of these; I will attempt to develop systematically the characteristic settings of a number of deposit types in terms of their positions at or near the margins of, or within, the lithospheric plates and secondarily to discuss how these may bear on some genetic problems.

The major features of the theory are reviewed briefly to set the scene (figures 1 and 2). At present, the surface of the globe consists largely of about eight major plates which are growing (accreting) along rises, for the most part near the midlines of the oceans, and moving symmetrically away. The plates are up to 150 kilometers thick; most have areas of both continental crust and much thinner oceanic crust overlying upper-mantle material. The plates rest on and move over the asthenosphere, a zone of little or no strength revealed by seismic data. Because the area of the earth remains essentially constant, additions to the plates at accreting margins must be balanced at the converging (consuming) margins; this is accomplished by descent of one plate beneath the other along a subduction zone marked by pronounced seismic activity (the Benioff zone). Two situations are illustrated in figure 2, collision of oceanic crust with a continent to produce a Cordilleran margin (the Andes) and with a small ocean basin to produce an island arc (Japan). Recent work *(Karig,* 1971) suggests that some arcs are moving away from the neighboring continents; thus a better example of an ocean/ocean confrontation would be the Mariana Arc, at present largely submerged.

The descending lithosphere partially fuses to form magmas whose composition varies with depth; at about 100 to 200 km these may have the calc-alkaline character *(Dewey* & *Bird,* 1970) with which many of the common hypogene ore deposits are associated *(Fonteilles,* 1967). These magmas rise to form extrusive and intrusive masses. Where two continental plates collide, as in the Himalayas, neither plate can descend, the force is dissipated in severe crushing with little or no magma generation, and endogenic ore deposits do not form.

Transform movement occurs where plates slide past each other without addition or subtraction as is taking place along most of the margin between North America and the Pacific Ocean today.

The present plates are considered to have developed by splitting of a single landmass (Pangaea) or of two landmasses (Laurasia and Gondwana) in post-Paleozoic time. Only a few „modern" splits of continental masses are shown on figure 1. — the Red Sea-Gulf of Aden and the Gulf of California — but many more, both active and inactive, are known or suspected (see, e. g. *Burke* & *Whiteman,* 1972; *Grant,* 1971, with respect to Africa).

In the simplest global tectonic terms, ore deposits can be thought as formed:
1. At or near plate margins
   a) Accreting (diverging)
   b) Transform
   c) Consuming (converging)

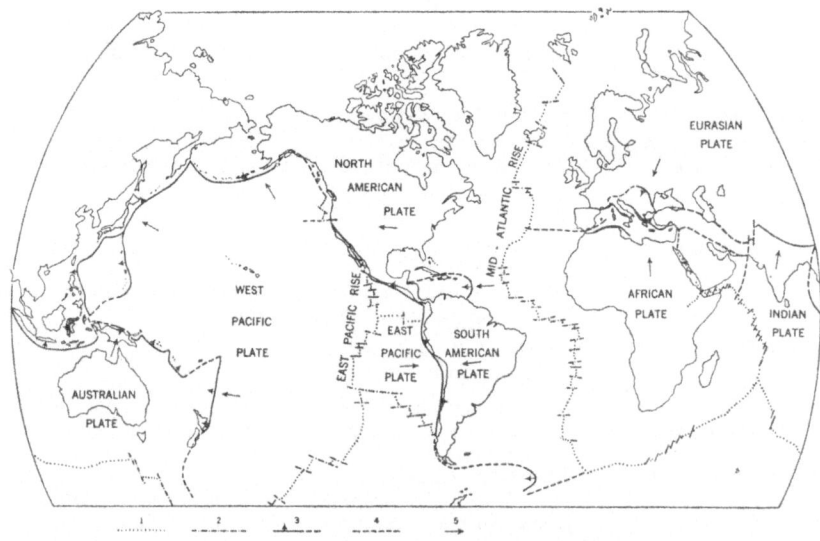

Fig. 1. Major lithospheric plates. 1) Accreting plate margin; 2) transform plate margin; 3) consuming plate margin with dip direction of downgoing plate; 4) margin of uncertain nature and (or) location; 5) relative plate motion.

2. Within plates
    a) In oceanic parts
    b) At trailing continental margins
    c) In continental parts

This oversimplification can serve as a starting point from which to evaluate the principal facts of the distribution of many types of ore deposits. The plate tectonic positions of some are clear, of others doubtful or nonspecific (two or more positions are possible), and of still others the conjunction of two factors seems to be required.

Table 1 (modified from *Guild*, 1971) expands the brief classification above with examples and draws attention to significant difference in the shapes and orientation of deposits, districts, and provinces. Those associated with plate margins tend to be elongated and to parallel the margins; furthermore, mineralization is roughly of the same age as the host rocks, ranging from syngenetic(?) to perhaps the close of a succeeding orogeny. By contrast, many deposits and districts formed within plates tend to be equidimensional; epigenetic mineralization may occur in far older host rocks; and provinces can cross the depositional-structural grain of a region at any angle. The remainder of this paper will be essentially an annotation of the table and discussion of some of its implications.

Table 1

Proposed relationship of some ore-deposit types to lithospheric plates

Deposits formed	Types, possible examples
1. *At or near plate margins*	Orientation of deposits, districts, and provinces tends to parallel margin
a) Accreting (diverging)	Red Sea muds. Ancient analogs? Certain cupriferous-pyrite (massive sulfide) ores, Cyprus? Newfoundland? Podiform Cr (may be carried across ocean and incorporated in island arc or continental margin)
b) Transform	Podiform Cr, Guatemala? Cu and Mn, Boleo, Baja California
c) Consuming (converging)	Chiefly of continent/ocean or island arc/ocean type; deposits formed at varying distances on side opposite oceanic, descending plate Podiform Cr, Alaska FeS₂-Cu-Zn-Pb stratabound massive sulfide, New Brunswick, Japan (Kuroko ores), California, British Columbia Mn of volcanogen type associated with marine sediments, Cuba, California, Japan Magnetite-chalcopyrite skarn ores, Puerto Rico, Hispaniola, Cuba, Mexico, California, British Columbia, Alaska Cu (Mo) porphyries, Puerto Rico, Panama, SW United States, British Columbia, Philippine Is., Bougainville Ag-Pb-Zn, Mexico, western United States, Canada Au, Mother Lode, California; Juneau Belt, Alaska Bonanza Au-Ag, western United States W, Sn, Hg, Sb, western North and South America
2. *Within plates*	Deposits tend to be equidimensional, distribution of districts and provinces less oriented (may be along transverse lineaments)
a) In oceanic parts	Mn-Fe (Cu, Ni, Co) nodules Mn-Fe sediments in small ocean basins with abundant volcanic contributions? Evaporites in newly opened or small ocean basins
b) At continental margins of Atlantic (trailing) type	Black sands, Ti, Zr, magnetite, etc. Phosphorite on shelf
c) In continental parts	Au (U) conglomerates, Witwatersrand Mesabi and Clinton types of iron formation Evaporites, Michigan Basin, Permian Basin; salt, potash, gypsum, sulfur Red-bed Cu; Kupferschiefer and Katangan Cu-Co U, U-V deposits, Colorado Plateau Fe-Ti-(V) in massif anorthosite, Canada, U.S. Stratiform Cr, Fe-Ti-V, Cu-Ni-Pt, Bushveld Complex Carbonatite-associated deposits of Nb, V, P, RE, Cu, F Kimberlite, diamonds Kiruna-type Fe-(P), SE Missouri Mississippi Valley type deposits, Pb-Zn-Ba-F-(Cu, Ni, Co)

Although the Red Sea metalliferous sediments *(Degens & Ross,* 1969) are the best known, least equivocal example of deposits formed at position 1a, accreting plate margins, certain of the stratabound massive sulfides, categorized in 1c (consuming margins) in table 1, are similar to them in both composition and form. Those pyrite and cupriferous pyrite deposits which occur in ophiolite sequences, generally believed to be generated at rises during ocean-floor spreading *(Bird & Dewey,* 1970), are also pleaced in 1a by a number of workers *(Hutchinson,* 1971; *Sillitoe,* 1972c; *Updahyay & Strong,* 1973). As the evidence for an island arc (1c) environment for many deposits such as Kuroko ores *(Tatsumi & Watanabe,* 1971) is incontrovertible, two plate positions are apparently possible for genesis of these stratabound massive sulfides. Gross distribution of these deposits (fig. 3) will not distinguish between those formed in 1a or 1c, except for the youngest, because the „conveyor belt" movement of the ocean floors toward subduction zones will cause deposits formed on accreting margins to end up near consuming margins eventually. The Paleozoic belts of the Urals (Variscan or Hercynian) and Scandinavia-Appalachia (Caledonian) probably mark former plate margins *(Guild,* 1973), but their nature at the time of ore deposition can only be determined from the sedimentary and petrologic features of the associated rocks.

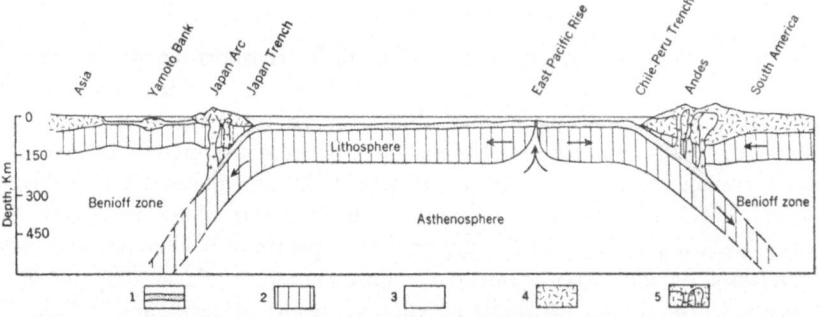

Fig. 2. Schematic section from South America to Asia illustrating major features of the plate tectonic theory. Not to scale. Based on *Dewey & Bird* (1970) 1) Oceanic crust; 2) upper mantle; 3) lower mantle; 4) continental crust; 5) calc-alkaline magmas, intrusive (+) and extrusive (v) products.

Chromite ores have several possible plate positions. Podiform masses in alpine peridotite are believed to have segregated in the mantle before introduction into the crust as already solidified masses carried up in crystal mushes *(Thayer,* 1942; *Guild,* 1947; Flint et al., 1948). Upwelling of the mantle peridotite, the depleted pyrolite of *Green & Ringwood* (1969) from which tholeiitic basalt magma has been abstracted, can introduce chromite at accreting margins. Obduction of

15

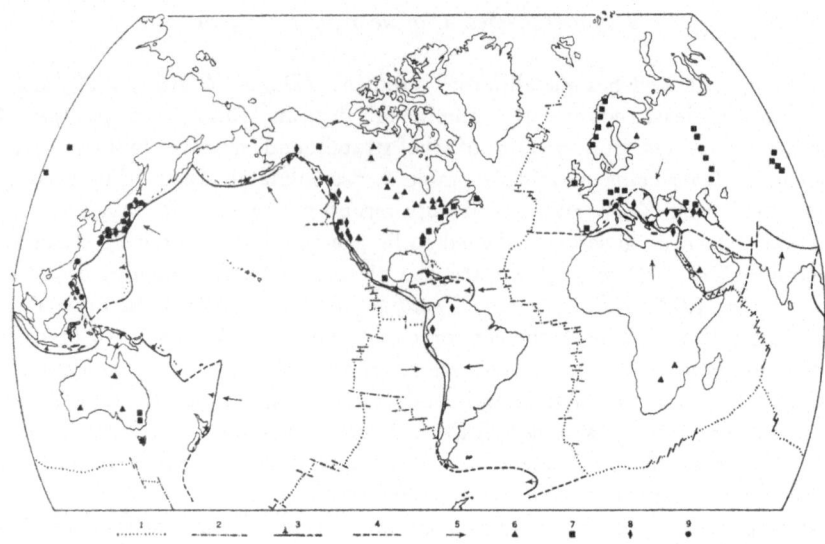

Fig. 3. Distribution of stratabound massive sulfide deposits. 1) Accreting plate margin; 2) transform plate margin; 3) consuming plate margin with dip direction of downgoing plate; 4) margin of uncertain nature and (or) location; 5) relative plate motion; 6) Precambrian deposit; 7) Paleozoic deposit; 8) Mesozoic deposit; 9) Cenozoic deposit.

ophiolite slices at consuming margins after their transport across an ocean floor by plate movement can emplace them in island arcs or continental margins. I believe that intrusion of peridotites along convergent margins by rheomorphic action can bring chromite segregations directly from the mantle as in the small oval intrusions of the Kenai Penisula *(Guild,* 1942) and Baranof Island *(Guild* & *Balsley,* 1942) of Alaska. Yet another possibility is along transform faults. *Dewey* & *Bird* (1970, p. 2630) suggest that serpentinite is injected along a zone of cataclasis on the active segment; chromite deposits could accompany this serpentinite. Chromiferous peridotite in lenses along the Motagua and Polochic fault zones in Guatemala may have had this origin. However, it has been suggested *(Newcomb,* 1973) that the Motagua fault zone is a Paleozoic suture; if so, the chromite deposits along it would have positions analagous to those in the Zagros crush zone of Iran.

Guilbert (1971) suggested that the copper and manganese deposits at Boleo on the west shore of the Gulf of California formed near or on a transform fault (position 1b). I subsequently speculated that the mineralization occurred at the intersection of the East Pacific Rise itself and an incipient transform fault *(Guild,* 1972a, p. 23), which would make the site a combination of 1a and 1b. It now seems probable to me that additional examples of such conjunctions (hybrid positions) will turn up as we learn more about the details.

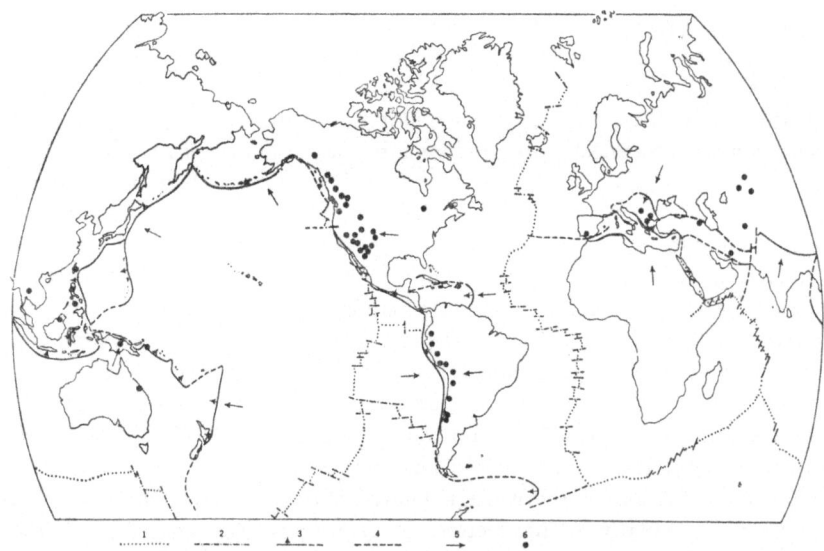

Fig. 4. Distribution of porphyry copper and molybdenum deposits. 1) Accreting plate margin; 2) transform plate margin; 3) consuming plate margin with dip direction of downgoing plate; 4) margin of uncertain nature and (or) location; 5) relative plate motion; 6) porphyry deposit.

Turning to the consuming margins, 1c, we see that many of the most common ore types have this position. In explanation, it seems evident that the energy to concentrate elements comes from forces activated by plate collision (magma generation, metamorphism, etc.). However, the elements (chiefly metals) may come from the downgoing oceanic plate, the overriding continental plate, or both. *Sillitoe* (1972b) has marshalled impressive evidence to support derivation of the porphyry copper deposits from the oceanic plate. Figure 4 shows that most porphyry deposits (both copper and molybdenum) are near the boundaries of the present plates and on the continent or island-arc side above the subduction zones. The simplest, most clearcut examples are along the Andean chain in South America and in the arcs of the Western Pacific. Elsewhere, particularly in the Southwestern Province of North America, the picture is not so clear. Instead of linear patterns paralleling closely plate margins, the deposits form a more or less equidimensional cluster 1200 by 1500 km across, in which lineaments or other intraplate structures seem to have exerted a major control on their distribution (*Tweto* & *Sims*, 1963, *Titley*, 1970; *Lowell*, 1973). The width of the zone and extended time span (Jurassic to Miocene) seem to preclude a simple relation to a Benioff zone. These deposits, and perhaps the Siberian porphyry deposits as well, may better be classed with those of „reactivated platforms" (*Guild*, 1972b), and hence fit more closely category 2c than 1c.

17

A number of other deposit types shown as 1c in table 1 may also be hybrid in the sense that they display characteristics of both plate margins and continental plate interiors.

Zonation of the base metals perpendicular to continental margins is a common phenomenon. In general, copper predominates near the margins in the eugeosynclinal parts of orogens, whereas lead and zinc are localized in the miogeosynclines and on platforms. *Laznicka & Wilson* (1972) have quantified this relationship using data from 4500 deposits throughout the entire world; their copper-lead lines (limits between copper-rich and lead-rich provinces) of Mesozoic and Tertiary mineralization lie near the plate margins. However, they note *(Wilson & Laznicka, 1972, p. 45)* that in the western United States three isolated major deposits, Butte, Bingham, and Ely, lie within the Paleozoic and Mesozoic miogeosynclinal domain. Lead isotope studies (e. g., *Murthy & Patterson*, 1961; *Stacey, Zartman & NKoma*, 1968, and *Zartman & Stacey*, 1971) indicate derivation of part of the lead from crustal rocks (because of its two-stage or multistage nature). Many districts of the western United States are localized on transverse structures and especially at intersections (e. g. *Billingsley & Locke*, 1941; *Mayo*, 1958; *Wisser*, 1959; *Jerome & Cook*, 1967; *Landwehr*, 1968). The upper lithospheric plate seems to have controlled the distribution of the deposits, and probably was the source for at least part of their metals as well.

Ores of lithophile elements such as tungsten and tin may be other examples of metals with hybrid plate positions. The distribution of tungsten deposits in the western United States (see e. g., *Kerr*, 1946, fig. 1) suggests that factors other than proximity to, or distance from, a subduction zone were operative in their genesis. The western limit of the deposits coincides approximately with the quartz diorite line *(Moore, 1959; Moore & al., 1963)*, hence they might be thought of as another example of control of chemical composition by depth of magma generation *(Kuno, 1959; Dickinson, 1968)*. It seems more likely, however, that the tungsten was derived from crustal materials by palingenesis *(Smirnov, 1968)* than from descending oceanic floor. The long-lived tin belts *(Schuiling, 1967; Petrascheck, 1968)* also indicate that certain areas have remained anomalously high in this element and that rejuvenation has played an important role, as *Schneider-Scherbina* (1964) has advocated for the tin deposits of Bolivia.

Deposits within plates

Many of the deposits formed within plates are exogenic and do not require extensive comment here. The nodules of the ocean floor are well known. Iron-manganese formations such as that in Aroostook County, Maine *(Pavlides, 1962)* and western New Brunswick were deposited at a time — Late Ordovician-Early Silurian(?) — when a contracting small ocean basin may have existed between an island arc and the mainland *(Bird & Dewey, 1970, p. 1049)*. Numerous eva-

18

porite sequences have formed in newly opening rifts (South Atlantic, Red Sea, Gulf of California) because of the restricted circulation of oceanic water. They could be considered as deposits at site 1a but because of their exogenic nature are placed here.

The products of tholeiitic magmas within the oceanic parts of plates (Hawaiian islands, etc.) are noteworthy for their lack of endogenic mineral deposits, and the likelihood that any conventional ore deposits were formed in position 2a seems remote.

The trailing continental margins (2b) are logical and obvious loci for heavy-mineral accumulations. The beach placers of rutile and (or) ilmenite, plus zircon, on the fragmented margins of Gondwana are outstanding examples. Source rocks for the titanium minerals, especially the rutile, are probably areas of granulite-facies metamorphism in the hinterland (Eric R. *Force*, U. S. Geol. Survey, oral communication, 1972). These are examples of a two- or multistage history in which ore genesis begins in one plate environment and subsequently is completed in another. Such deposits contrast with the hybrid type that combines features of two plate positions simultaneously. Conversion of a passive margin with heavy-mineral concentrations to an active one *(Dewey & Bird, 1970, p. 2638—2640)* could introduce a third stage of ore concentration through palingenesis of low-grade clastic terrigenous materials.

Very diverse types of deposits are formed in 2c, the continental parts of the lithospheric plates. Here are the major deposits of sedimentary, exogenic ores, both the Proterozoic iron formations and the Phanerozoic ironstones, and also the evaporites of the large basins. Redbed copper and the somewhat related uranium and uranium-vanadium deposits of the Colorado Plateau and elsewhere form here, as do the Kupferschiefer and Katangan copper deposits, whatever their precise genesis may be.

Endogenic deposits in continental settings form another major group that comprises both those with more or less close association with igneous rocks of widely different compositions and those with no, or at best equivocal relationship to any igneous activity. The igneous-associated deposits include the stratiform chromite deposits (Bushveld, Stillwater, et al.) and the nickel-copper-platinum and the vanadium-bearing titaniferous magnetite layers best exemplified in the Bushveld. The immense size, uniformity of composition and structure, and relationship of the Bushveld Complex to the subjacent sedimentary rocks attest to intrusion of mafic magma from a deep source (lower crust or mantle) into a stable platform.

The nonlayered iron-titanium deposits, also moderately vanadiferous, that occur in the massif-type anorthosite bodies in many parts of the world, and iron-(apatite) and iron-copper deposits in Missouri associated with alkali-rich extrusive and intrusive rocks are other example of deposits introduced into continental lithosphere from deep sources.

Deposits of 2c discussed thus far are not numerous enough to have been generally related to lineaments (although such major examples as the Great Dyke and

Muskox indicate that the magma arose along extensive fractures that must have penetrated the sialic crust), but the distribution of carbonatite-associated deposits of niobium, vanadium, phosphate, and rare earths shows definite linear patterns that are not related to orogenic belts. This is also true for the kimberlites. Certain magnetite, titanium, and barite deposits are among the less common but in places very large deposits that also have this general lithospheric plate setting. Examples are known on all the continents, and they span much of the geologic record since early Precambrian time. In the southern hemisere many may be related to the Mesozoic breakup of Gondwana, but others are probably associated with abortive fracturing that did not lead to actual rifting *(Burke & Whiteman, 1972).*

Prime examples of deposits with little or no apparent relationship to magmatic rocks are the so-called Mississippi Valley type of lead-zinc-(copper)-barite-fluorite ores, which are, in North America at least, obviously in 2c. Without going into the details, which have been thoroughly documented in a symposium organized by Charles Behre *(Brown, 1967),* many are located in carbonate platform rocks overlying old shield areas. The deposits in any one district have a pronounced tendency to be restricted to one or a few favorable stratigraphic units; various local features (reefs, pinchouts, solution breccias, etc.) determine the precise depositional environment. However, on a continental scale the North American districts occur along lineaments which are also the loci of alkaline intrusive rocks, including kimberlites, and of cryptoexplosion structures. The best documented of these, the 38th parallel lineament, has been traced some 1300 km in a westerly direction from the Appalachians to the mid-continent and may extend at least another 900 km to the Rocky Mountains *(Heyl, 1972). Snyder* (1970) lists 8 mineral districts, 10 igneous features, and 8 explosion events along or near it and draws attention to similar features, the Tennessee lineament near the 36th parallel and the Galena lineament near the 42nd, that are also loci of major districts together with igneous and explosion phenomena. In Canada, the Pine Point district directly overlies the McDonald fault that divides the Slave and Churchill Provinces of the Shield and has been traced in the subsurface many hundreds of kilometers to the southwest across the Phanerozoic Interior Lowlands. All these structures must have been long-lived; intermittent activity along the 38th parallel lineament extended from at least Cambrian to Tertiary time, and seismic events are continuing at present. It seems evident, and the inclusions in the kimberlites prove *(Brookins, 1969),* that the fracture zone penetrated the sial at times, but the source of the metals in the ore deposits was not necessarily in the mantle. Lead isotope data indicate that at least some of the lead was derived from crustal rocks and it may well be, as *Snyder* (1970, p. 91) and others have suggested, that heat rising along lineament zones has increased the geothermal gradient and provided the energy to concentrate metals from the connate brines in deep basins. We probably have another example of multistage ore concentration. Whether the North American pattern is applicable to that in the Eastern Hemisphere is a question I am prepared to answer, but the similarities of the deposits described by numerous European colleagues *(Brown, 1967)* to those of the Mississippi Valley districts are so close in many respects that some common mode of origin seems inevitable.

A relatively unimportant district may provide a clue to the relationship between platforms, cover, and deep fractures in metallogenesis. Lead-zinc-(copper-fluorite) deposits in the Benue trough of Nigeria *(Farrington, 1952)* occur in Cretaceous sedimentary and volcanic rocks that filled a rift (aulacogen) that is believed to have opened at the triple junction formed as South America and Africa began to separate *(Grant, 1971)*. Though most are veins in clastic or volcanic rocks, some apparently are stratiform in limestone; the mineralogy is, in any case, reminiscent of the Mississippi Valley type under discussion. *Dewey & Burke* (1973) have recently suggested that such structures (including this one) form in response to uplift of crustal lithosphere over mantle plumes. Such plumes would provide the relatively low but persistent heat indicated by fluid-inclusion studies *(Roedder, 1971)* to have been present during a prolonged period in districts in the eastern United States. *Macintyre* (1973) has suggested that plumes in continental settings are responsible for emplacement of nephelinitic, carbonatitic, and kimberlitic rocks, and that periodic pluming may have reactivated major lineaments and rifts. This new concept of deep vertical movement and consequent upward propagation of energy from subcrustal sources, with or without magma transfer, may provide the mechanism called for years ago by *Billingsley & Locke* (1941, p. 59) to account for many of the ore districts of the United States. It also can explain the relationship between doming and ore districts documented by *Wisser* (1960).

Conclusion

If the Benue trough and, by extrapolation, the Mississippi Valley deposits are in any way rift associated, we seem to have come nearly full circle, except that the deposits in 2c are in a continental setting whereas those in 1a have an oceanic one. Both seem to derive their primary energy from vertical, deep-seated sources, but the ore-forming elements of the former may have been concentrated from disperse crustal reservoirs in contrast to a supposed mantle source for those in 1a. Similarly, the deposits tentatively assigned to 1c but noted as transitional to those of platforms have many similarities to deposits discussed under 2c. From the plate-tectonic viewpoint I suggest that the critical distinction here is whether subduction and regeneration of elements present in the downgoing slab played a dominant role in ore formation or whether energy (heat) was the principal or only factor directly related to plate convergence.

The tentative classification of table 1 is far from definitive; for example, many types of deposits are not mentioned. Obviously more facts are needed concerning the environments of deposits and the relationship between various categories, and, of course, both the validity of the plate tectonic concept and its details require confirmation through additional research. However, I believe that we are in a position to define some fundamental problems of ore genesis and hence to design studies directed toward their solution. We should also be able before too much longer to explain the distribution of the so-called metallogenic provinces in rational terms that are consonant with the major aspects of earth science that seem to be evolving so rapidly.

References

American Geol. Inst. (1972): Glossary of Geology, Washington, D. C. 805 + 52 p.

Billingsley, P. & Locke, A. (1941): Structure of ore districts in the continental framework. — Am. Inst. Min. and Met. Engrs. Trans., 144, p. 9—64.

Bird, J. M. & Dewey, J. F. (1970): Lithosphere plate-continental margin tectonics and the evolution of the Appalachian orogen. — Geol. Soc. America Bull., 81, p. 1031—1060.

Brookins, D. G. (1969): Riley County, Kansas kimberlites and their inclusions (abs.). — Geol. Soc. America Abs. with Programs for 1969, pt. 2, p. 4.

Brown, J. S. ed. (1967): Genesis of stratiform lead-zinc-barite-fluorite deposits (Mississippi Valley type deposits). — A symposium, New York, 1966: Econ. Geology Mon. 3, 443 p.

Burke, K. & Whiteman, A. J., 1972, Uplift, rifting, and the breakup of Africa (abs.). — Am. Geophys. Union Trans., 53, p. 517.

Degens, E. T. & Ross, D. A., eds. (1969): Hot brines and Recent heavy metal deposits in the Red Sea. — Springer-Verlag, Inc., New York, 600 p.

Dewey, J. F. & Bird, J. M. (1970): Mountain belts and the new global tectonics. — Jour. Geophys. Research, 75, p. 2625—2647.

Dewey, J. F. & Burke, K. (1973): Plume generated triple junctions (abs.). — Am. Geophys. Union Trans., 54, p. 239.

Dickinson, W. R. (1968): Circum-Pacific andesite types. — Jour. Geophys. Research, 73, p. 2261—2269.

Farrington, J. L. (1952): A preliminary description of the Nigerian lead-zinc field. — Econ. Geology. 47, p. 583—608.

Flint, D. E., de Albear, J. F. & Guild, P. W. (1948): Geology and chromite deposits of the Camagüey district, Camagüey Province, Cuba. — U.S. Geol. Survey Bull. 954-B, p. 39—63.

Fonteilles, M. (1967): Appréciation de l'intérêt metallogénique du volcanisme de Madagascar à partir de ses caractères pétrologiques. — France, Bur. Recherches Géol. et Minères Bull. No. 1, p. 121—154.

Grant, N. K. (1971): South Atlantic, Benue trough, and Gulf of Guinea Cretaceous triple junction. — Geol. Soc. America Bull., 82, p. 2295—2298.

Green, D. H. & Ringwood, A. E. (1969): The origin of basaltic magmas. — In: P. J. Hart, ed., The Earth's crust and upper mantle, Am. Geophys. Union Geophys. Mon. 13, p. 489—495.

Guilbert, J. M. (1971): Known interactions of tectonics and ore deposits in the context of new global tectonics. — Am. Inst. Min. Metall. Petrol. Engrs., Soc. Mining Engrs. Preprint 71-S-91, 19 p.

Guild, P. W. (1942): Chromite deposits of Kenai Peninsula, Alaska. — U.S. Geol. Survey Bull. 931-G, p. 139—175.

— (1947): Petrology and structure of the Moa chromite district, Oriente Province, Cuba. — Am. Geophys. Union Trans., 28, p. 218—246.

Guild, P. W. (1971): Metallogeny: a key to exploration. — Mining Eng., 23, no. 1, p. 69—72.

— (1972a): Metallogeny and the new global tectonics. — Internat. Geol. Cong., 24th, Montreal, Repts., sec. *4*, p. 17—24.

— (1972b): Massive sulfides vs. porphyry deposits in their global tectonic settings (abs.). — Jour. Mining Met. Inst. Japan, *88*, p. 725.

— (1973): Massive sulfide deposits as indicators of former plate boundaries (abs.). — Econ. Geology, *68*, p. 137—138; U.S. Geol. Survey Open-file report, 11 p.

Guild, P. W. & *Balsley*, J. R., *Jr.* (1942): Chromite deposits of Red Bluff Bay and vicinity, Baranof Island, Alaska. — U.S. Geol. Survey Bull. *936-G*, p. 171—187.

Heyl, A. V. (1972): The 38th parallel lineament and its relationship to ore deposits. — Econ. Geology *67*, p. 879—894.

Hutchinson, R. W. (1971): Volcanogen sulfide deposits and their metallogenic significance (abs.). — Mining Eng., *23*, no. 12, p. 71.

Jerome, S. E. & *Cook*, D. R. (1967): Relation of some metal mining districts in the western United States to regional tectonic environments and igneous activity. — Nevada Bur. Mines Bull. *69*, 35 p.

Karig, D. E. (1971): Origin and development of marginal basins in the western Pacific. — Jour. Geophys. Research, *76*, p. 2542—2561.

Kerr, P. F. (1946): Tungsten mineralization in the United States. — Geol. Soc. America Mem. *15*, 241 p.

Kuno, H. (1959): Origin of Cenozoic petrographic provinces of Japan and surrounding areas. — Bull. Volcanol., ser. 2, *20*, p. 37—76.

Landwehr, W. R. (1968): The genesis and distribution of major mineralization in western United States. — Econ. Geology, *63*, p. 967—970.

Laznicka, P. & *Wilson*, H. D. B. (1972): The significance of a copper-lead line in metallogeny. — Internat. Geol. Cong., 24th, Montreal, Repts., sec. *4*, p. 25—36.

Lowell, J. D. (1973): Regional characteristics of southwestern North America porphyry copper deposits. — Am. Inst. Min. Metall. Petrol. Engrs., Soc. Mining Engrs., Preprint *73-S-12*, 37 p.

Macintyre, R. R. (1973): Possible periodic pluming (abs.). — Am. Geophys. Union Trans., *54*, p. 239.

Mayo, E. B. (1958): Lineament tectonics and some ore districts of the southwest. — Mining Eng., *10*, p. 1169—1175.

Mitchell, A. H. & *Garson*, M. S. (1972): Relationship of porphyry copper and circum-Pacific tin deposits to paleo-Benioff zones. — Inst. Mining and Metallurgy Trans. B., *81*, no. 783, p. B10—B25.

Moore, J. G. (1959): The quartz diorite boundary line in the western United States. — Jour. Geology, *67*, p. 198—210.

Moore, J. G., *Grantz*, A. & *Blake*, M. C., *Jr.* (1963): The quartz diorite line in northwestern North America. — U.S. Geol. Survey Prof. Paper *450-E*, p. E89—E93.

Murthy, V. R. & *Patterson*, C. (1961): Lead isotopes in ores and rocks of Butte, Montana. — Econ. Geology, *56*, p. 59—67.

Newcomb, W. E. (1973): Late Paleozoic tectonics of the Motagua fault: a subduction-collision scar? (abs.). — Am. Geophys. Union Trans., *54*, p. 471—472.

Pavlides, L. (1962): Geology and manganese deposits of the Maple and Hovey Mountains area, Aroostook County, Mainer. — U.S. Geol. Survey Prof. Paper *362*, 116 p.

Petrascheck, W. E. (1968): Kontinentalverschiebung und Erzprovinzen. — Mineralium Deposita, *3*, p. 56—65.

Roedder, E. (1971): Fluid-inclusion evidence on the environment of formation of mineral deposits of the Southern Appalachian Valley. — Econ. Geology, *66*, p. 777—791.

Sawkins, F. J. (192): Sulfide ore deposits in relation to plate tectonics. — Jour. Geology, *80*, p. 377—397.

Schneider-Scherbina, A. (1964): Epocas y zonas metalogenéticas. — In *Ahlfeld* Federico, and *Schneider-Scherbina*, A., Los yacimientos minerales y de hidrocarburos de Bolivia: Bolivia Min. de Minas y Petroleo, Bol. 5, 188 p., p. 31—39.

Schuiling, R. D. (1967): Tin belts on the continents around the Atlantic Ocean. — Econ. Geology, *62*, p. 540—550.

Sillitoe, R. H. (1972a): Relations of metal provinces in Western America to subduction of oceanic lithosphere. — Geol. Soc. America Bull., *83*, p. 313—318.

— (1972b): A plate tectonic model for the origin of porphyry copper deposits. — Econ. Geology, *67*, p. 184—197.

Sillitoe, R. H. (1972c): Formation of certain massive sulphide deposits at sites of sea floor spreading. — Inst. Mining and Metallurgy Trans. B., *81*, no. 789, p. B141—B148.

Smirnov, V. I. (1958): The sources of ore-forming material. — Econ. Geology, *63*, p. 380—389.

Snyder, F. G. (1970): Structural lineaments and mineral deposits, eastern United States. — In AIME World Symposium on Mining and Metallurgy of Lead and Zinc, v. 1, Mining and concentrating of lead and zinc. — Am. Inst. Min. Met. and Petrol. Engrs., New York, p. 76—94.

Stacey, J. S., *Zartman*, R. E. & *NKomo*, I. T. (1968): A lead isotope study of galenas and selected feldspars from mining districts in Utah. — Econ. Geology, *63*, p. 796—814.

Tatsumi, T. & *Watanabe*, T. (1971): Geological environment of formation of the Kuroko-type deposits. — Soc. Mining Geol. Japan, Spec. Issue *3*, p. 216—220. (Proc. IMA-IAGOD Meetings '70, IAGOD Vol.).

Thayer, T. P. (1942): Chrome resources of Cuba. — U.S. Geol. Survey Bull. *935-A*, p. 1—74.

Titley, S. R. (1970): Paleotectonic environment of Arizona porphyry copper deposits (abs.). — Geol. Soc. America Abs. with Programs, 2, no. 7, p. 707—708.

Tweto, O. & *Sims*, P. K. (1963): Precambrian ancestry of the Colorado mineral belt. — Geol. Soc. America Bull. *74*, p. 991—1014.

Updahyay, H. D. & *Strong*, D. F. (1973): Geological setting of the Betts Cove copper deposits, Newfoundland: an example of ophiolite sulfide mineralization. — Econ. Geology, *68*, p. 161—167.

Wilson, H. D. B. & *Laznicka*, P. (1972): Copper belts, lead belts, and copper-lead lines of the world. — Internat. Geol. Congress., 24th, Montreal, Repts., sec. *4*, p. 37—51.

Wisser, E. H. (1959): Cordilleran ore districts in relation to regional structure. — Canadian Inst. Mining Metall., Bull. *561*, p. 34—42.

— (1960): Relation of ore deposition to doming in the North American Cordillera. — Geol. Soc. America Mem. 77, 117 p.

Zartman, R. E. & *Stacey*, J. S. (1971): Lead isotopes and mineralization ages in Belt Supergroup rocks, northwestern Montana and northern Idaho. — Econ. Geology, *66*, p. 849—860.

The Non-relation Between Metal Provinces and Theories of Plate Tectonics in the Western United States

James A. *Noble**

Zusammenfassung

„Über das Fehlen einer Beziehung zwischen Erzprovinzen und der Theorie der Plattentektonik in den westlichen Vereinigten Staaten"

In einer früheren Abhandlung *(Noble,* 1970) wurde der Schluß gezogen, daß die Anordnung und Verteilung der Erzprovinzen in den westlichen Vereinigten Staaten das Ergebnis einer ursprünglichen, vor-krustalen Heterogentität im Oberen Mantel sei, und daß die Metalle dieser Lagerstätten aus dem Oberen Mantel herzuleiten seien. Die enge räumliche und zeitliche Verknüpfung zwischen den meisten Erzlagerstätten und einigen Intrusivgesteinen wurde strukturell bedingt gedeutet: der Aufstieg von Magma aus dem Oberen Mantel ließ druckentlastende Kanäle entstehen, die den Aufstieg von Metallen und anderen erzbildenden Bestandteilen erleichterten.

Eine wachsende Zahl von Altersdatierungen aus den westlichen Vereinigten Staaten erlauben es nun, die Altersverteilung der extrusiven und intrusiven Gesteine sowie der Erzlagerstätten in Karten zusammenzuzeichnen. Diese Altersverteilungen zeigen im allgemeinen konzentrische Anordnung, wobei die älteren Einheiten außen, die jüngeren im Zentrum liegen. Die Art der Anordnung scheint wegen der großen Menge an abgesetzten Metallen, wegen dem Alter der Lagerstätten und der sehr großen Verbreitung der Erzprovinzen mit keinem Modell der Plattentektonik vereinbar zu sein. Es wird daher der Schluß gezogen, daß die Erzlagerstätten der westlichen Vereinigten Staaten durch die „new global tectonics" nicht berührt wurden.

* James A. *Noble,* Consultant Mining Geology, 1475 East California Blrd., Pasadena/ Calif., USA.

Abstract

In a previous paper *(Noble, 1970)*, it was concluded that the patterns of distribution of metal provinces in the western United States were a result of a primitive or precrustal heterogeneity in the upper mantle, and that the metals of the ore deposits making up these patterns were derived from the upper mantle. The close space and time association between most ore deposits and some intrusive rocks was believed to be structural: the ascent of magma from the upper mantle provided avenues of released pressure that facilitated the rise of metals and other ore-forming constituents.

Models derived from theories of plate tectonics have been developed to explain tectonism, the development of extrusive and intrusive rocks, and the formation of ore deposits *(Coats, 1962; Hamilton, 1969; Dewey & Bird, 1970; Dietz, 1970; James, 1971; Sillitoe, 1972a, b; Clark & Zentilli, 1972)*. In particular, subduction zones developed by the collision of oceanic and continental plates have been postulated, but ascent of molten material from the mantle at the boundaries between moving plates has also been suggested. I will discuss the space and time relationships between extrusion, intrusion, and ore formation in the western United States in terms of possible subduction zones.

The extrusive rocks can be divided into three age groups, Laramide, mid-Tertiary, and Pliocene-Pleistocene. The Laramide pattern forms a U-shape, continuing NNW into Canada, with a maximum spread of 800 mi (1280 km). The mid-Tertiary rocks occupy a large area mostly inside the Laramide area. The Pliocene-Pleistocene rocks, predominantly basalts and andesites, form one belt of andesite near the coast and two transverse belts of basalt, all overlapping the earlier rocks. The subduction hypothesis would require two subduction zones in opposite directions, dipping inward, a most unlikely situation. The distribution of the Pliocene-Pleistocene rocks might mark boundaries between moving plates, an ad hoc assumption at present.

The intrusive rocks have not been adequately dated, and their pattern can be discussed only in general terms. Large Mesozoic batholiths predominate near the coast and extend in places as much as 500 mi (800 km) eastward. Smaller Laramide batholiths and plutons are very extensive. For the most part they lie to the east of the Mesozoic batholiths, to a maximum distance of 1200 mi (1900 km) from the coast. The two age groups overlap somewhat along a central belt; more dating is needed here. Mid-Tertiary intrusive rocks are less extensive (inadequate dating?). The principal occurrence overlaps the boundaries between the older groups, but there is one linear belt near the coast. The space-time relationships between Mesozoic and Laramide groups would satisfy a subduction model only with a dip of a few degrees, and the mid-Tertiary group would not be explained.

The pattern of distribution of the ore deposits is more complex than those of the intrusive and extrusive rocks, in part because a greater proportion of the ore deposits has been dated by isotopic techniques. Mesozoic ore deposits form an arcuate belt parallel to the coast, mostly within 200 mi (320 km) of the coast but in places extending to 450 mi (720 km) inland. Laramide ore deposits, the most important in terms of total value, also form an arc, but this time facing away from the coast to a maximum distance of 1000 mi (1600 km). They continue into Canada but for only a short distance into Mexico. Tertiary deposits are somewhat less extensive but quite important commercially. They occupy the gap between the Mesozoic and Laramide arcs and also overlap the Laramide areas. Datings presently available permit separating late Tertiary from mid-Tertiary, but the areas of occurrence are similar, and additional sampling may close this gap. If we attempt to relate the pattern of ore deposition to a hypothesis of subduction, once again, as for the extrusive rocks, we need two opposite facing subduction zones.

Conclusion

These studies do not contradict the general conclusions of the hypotheses of continental drift, sea-floor spreading, and plate tectonics, but they indicate to me that probably the processes of extrusion and intrusion and certainly the process of ore formation in the western United States have no relation to those hypotheses and were not governed by them. The initiation of sea-floor spreading in the Atlantic Ocean and the first great surge of batholithic intrusion in western United States, both in early Mesozoic time, were nearly synchronous and may have been related, but the nature of the relationship is not clear. Some metallization accompanied this orogeny, but unless very large amounts have been completely destroyed by erosion, which seems unlikely, the succeeding Laramide orogeny and volcanism produced much greater amounts of metallization, without any clear time relationship to global tectonics. The simultaneous concentration and separation of metals, on an arcuate zone reaching to distances of about 1000 mi (1600 km) from the continental margin, cannot be fitted into any global tectonic model known to me. Large forces, or large amounts of heat, are required to explain the volcanism and metal formation in the western United States, and it seems likely that these were contained within the areas I have described; we need to look down, not to one side, for the sources (Gilluly, 1970). It has been noted (Damon & Mauger, 1966; Armstrong, 1970) that volcanism on both large and small scale tends to begin with a great surge and then taper off gradually. It has also been noted here that volcanism shows some tendency to contract toward a center; in a broad way, older units surround younger units in the age distribution maps. These two generalizations seem to point to the sporadic and rather sudden generation at great depths of large amounts of heat by mechanisms for which at present we probably have no adequate explanations. The rise and emplacement of the metals to form ore deposits is probably part of this process, and these likewise are not yet explained.

References

Armstrong, R. L. (1970): Geochronology of Tertiary igneous rocks, eastern Nevada and vicinity, USA. — Geochim. Cosmochim. Acta, *34*, p. 203—232.

Clark, A. H. & *Zentilli*, M. (1972): The evolution of a metallogenic province at a consuming plate margin: The Andes between latitudes 26⁰ and 29⁰ South. — Canadian Min. Met. Bull., *65*, p. 37 (abstr.).

Coats, R. R. (1962): Magma types and crustal structure in the Aleutian Arc. In *Macdonald*, G. A. & *Kuna*, H., Editors: The crust of the Pacific Basin: Amer. Geophys. Union Geophys. Mon. *6*, p. 92—109.

Damon, P. E. & *Mauger*, R. L. (1966): Epeirogeny-orogeny viewed from the Basin and Range Province. — Amer. Inst. Min. Eng. Trans., *235*, p. 99—111.

Dewey, J. F. & *Bird*, J. M. (1970): Mountain belts and the new global tectonics. — Jour. Geophys. Res., *75*, p. 2625—2647.

Dietz, R. S. (1970): Continents and ocean basins. — In *Johnson*, H. & *Smith*, B. L., Editors: The megatectonics of continents and oceans: New Brunswick, N. J., Rutgers Univ. Press, p. 24—46.

Gilluly, J. (1970): Crustal deformation in the western United States. — In *Johnson*, H. & *Smith*, B. L., Editors: The megatectonics of continents and oceans: New Brunswick, N. J., Rutgers Univ. Press, p. 47—73.

Hamilton, W. (1969): Mesozoic California and the underflow of Pacific mantle. — Geol. Soc. Amer. Bull., *80*, p. 2409—2430.

James, D. E. (1971): Plate tectonic model for the evolution of the central Andes. — Geol. Soc. Amer. Bull., *82*, p. 3325—3346.

Noble, J. A. (1970): Metal provinces of the western United States. — Geol. Soc. Amer. Bull., *81*, p. 1607—1624.

— (1974): Metal provinces and metal finding in the western United States. — Mineralium Deposita (in press).

Sillitoe, R. H. (1972a): A plate tectonic model for the origin of porphyry copper deposits. — Econ. Geol., *67*, p. 184—197.

— (1972b): Relation of metal provinces in western Amerika to subduction of oceanic lithosphere. — Geol. Soc. Amer. Bull., *83*, p. 813—817.

The substance of this paper was first presented at the Leoben symposium on November 6. 1972. A more extended discussion, with applications to ore finding, will appear in Mineralium Deposita (Noble, 1974).

Geochemie erzführender Provinzen in phanerozoischen Plattformen

Kingsley *Dunham**

Zusammenfassung

Einige Merkmale metallführender Provinzen, wie sie von W. E. *Petrascheck* (1965) unter der Bezeichnung „stabile Tafeln" und „Bedeckung der starren Tafeln" beschrieben wurden, werden untersucht. Typisch für die Ausbildung dieser Provinzen sind sedimentäre Abfolgen bis 3500 m mächtig. Sie liegen diskordant auf altem, stark gestörtem Grundgebirge. Bruchtektonik kann, muß aber nicht eine Rolle spielen. Eine einfache Folge von Elementen erreicht wirtschaftliche Konzentrationen in dieser Art von Vorkommen, besonders Cu, Zn, Pb, U, Ba, F. Diese Vorkommen sind vielleicht mit Permeabilitätskanälen, wie sie z. B. in Sandsteingrundwasserleitern oder verkarsteten Kalken zur Verfügung standen, oder mit regionalen Spaltensystemen in Beziehung zu setzen. Die Geochemie der auftretenden Elemente ist viel einfacher als die der Orogene oder der präkambrischen kratonischen Provinzen. Die Ablagerung von Sedimenten in epikontinentalen Meeren, die die Plattformen überfluteten, enthalten einige wirtschaftliche Vorkommen, besonders Fe und Cu.

Abstract

„Geochemistry of Metalliferous Provinces in Phanerozoic Platforms"

Some features of metalliferous provinces grouped by W. E. *Petrascheck* under the titles „Stabile Tafeln" and „Bedeckung der Starren Tafeln" (1965) are examined. The typical geological setting is in a sedimentary sequence up to 3500 m. thick, no more than gently folded, unconformably overlying highly distorted ancient basement rocks. Block faulting may or may not be important. A simple

* Ausländisches korrespondierendes Mitglied der Österreichischen Akademie der Wissenschaften, Sir Kingsley *Dunham*, Prof. Dr., Institute of Geological Sciences, Exhibition Road, South Kensington, London SW 7/2 DE, Great Britain.

29

suite of elements reaches economic concentrations in this setting, notably Cu, Zn, Pb, U, Ba, F; these may be related to permeability channels provided for example by sandstone aquifers or karst features in limestone; or to regional fissure systems. The geochemical assemblage of associated elements is noticeably simpler than in the mobile belt, or Pre-Cambrian cratonic provinces. Sediment deposition in the epicontinental seas which spread over the platforms included some economic concentrations, notably of Fe and Cu.

Metallogenetische Provinzen und globale Tektonik

Professor *Petrascheck* diskutierte in seinem Vortrag 1964 vor der Deutschen Akademie der Wissenschaften zu Berlin und in seiner anregenden Veröffentlichung (1965) über metallogenetische Provinzen unter anderem die Eigenschaften erzhaltiger Regionen, die in drei größeren, kontrastierenden, megatektonischen Einheiten auftreten: 1. in Orogenen, 2. in metamorphen Schilden, 3. in Plattformen, in denen Grundgebirge von kaum gestörten Sedimenten überlagert wird. Kontinente werden hier als eine Kombination, aufgebaut aus diesen drei Einheiten, betrachtet. Folgt man dem Konzept der Kontinentalverschiebung, das im letzten Jahrzehnt stark in den Vordergrund getreten ist, so liegen rezente Orogene anerkannterweise an Kontinentalrändern, wo ozeanische Kruste unter den Kontinent abtaucht oder eine kontinentale Platte eine andere überfährt. Obwohl tiefe, lang ausgezogene Tröge, die große Sedimentmächtigkeiten erreichen können (gemeint sind Geosynklinalen im Sinne *Dana-Hall)*, oftmals an Kontinentalränder gebunden sind, hat man nun erkannt, daß Orogenesen nicht allein an geosynklinale Tröge gebunden sind. Es ist klar, daß mobile Zonen und Orogene in ihrer aktiven Zeit Gebiete höchst intensiver physikochemischer Prozesse in der Kruste sind, was sich in intensiver Faltung und Metamorphose der Gesteine ausdrückt und von magmatischen Intrusionen und Extrusionen begleitet wird. *Petrascheck* weist darauf hin, daß die begleitenden metallogenetischen Provinzen wie die Orogengürtel lineare Erstreckung zu haben scheinen. Viele Autoren haben kürzlich darauf aufmerksam gemacht, wie wichtig der aktive Kontinentalrand in Beziehung zur Mineralisation ist (z. B. *Guild* 1971, *Mitchell* & *Garson* 1972, *Sillitoe* 1972, *Dunham* 1973).

Die metamorphen Schilde wurden weitgehend, wie viele annehmen, durch eine Aufeinanderfolge früherer Orogenesen gebildet, als deren Ergebnis metamorphisierte, granitisierte und intrudierte Gesteine zusammengeschweißt wurden und so die Basis der stabilen Plattformen bildeten. Die Entwicklung dieser Plattformen beinhaltet ebenfalls das Konzept tiefer Abtragung, die zur Peneplainisierung aufgrund mariner oder fluviatiler Einwirkung führte. Doch nicht alle präkambrischen Gebiete der Erdkruste unterlagen diesen Bedingungen. Sie formen heute große Gebiete mit niedrigem Relief wie z. B. in N-Kanada, Karelien und Sibirien, oder Hochplateaus wie in Afrika. Metallogenetische Provinzen in diesen Schilden haben normalerweise, wie *Petrascheck* ausführte, sehr unregelmäßige

Grenzen und sind sehr stark von den jeweiligen geologischen Gegebenheiten abhängig.

Zu verschiedenen Zeiten im Präkambrium und besonders im Proterozoikum erhielten die eingeebneten kristallinen Gesteine eine Sedimentbedeckung, die stellenweise kaum gestört ist. Das war im wesentlichen noch mehr der Fall im Phanerozoikum, als Gebiete, die sich mindestens über die Hälfte der Kontinente erstrecken, von epikontinentalen Meeren überflutet und mit Sedimenten bedeckt wurden, die, obwohl hauptsächlich im Flachwasserbereich abgelagert, Mächtigkeiten von 3000 m und mehr erreichten, wo die Absenkung entsprechend war. Professor *Petrascheck* sowie vor ihm *Blondel* (1937) lenkten unsere Aufmerksamkeit auf die *Bedeckung der starren Tafeln*. *Blondel* hob als erster die weitverbreitete aber relativ einfache Art der Mineralisation solcher Gebiete hervor; und *Petrascheck* bemerkte, daß die metallogenetischen Provinzen in dieser Art von Ausbildung, allgemein gesprochen, lateral extensiv sind, ganz im Gegensatz zur linearen Ausbildung in den Orogenen.

Mir scheint, daß einige Gefahr darin liegt, daß man vom metallogenetischen Gesichtspunkt dem mobilen oder orogenen Typ der Mineralprovinzen oder ihren metamorphisierten Äquivalenten in den alten Schilden zu große Bedeutung beimißt im Vergleich zu denen in bedeckten Plattformen. Niemand bezweifelt natürlich die große Bedeutung der aktiven Kontinentalränder in der Metallogenese; hier findet man die weitest verbreiteten und höchst komplizierten Lagerstätten der Grundmetalle und einiger wertvoller Metalle und im besonderen die der Porphyrkupfererze. *Kelly* & *Turneaure's* Beschreibung (1970) der Mineralisation der östlichen bolivianischen Anden gibt die Kompliziertheit im mineralogischen Bereich recht gut wider. Die meisten der Lagerstätten stehen in mehr oder weniger enger Beziehung zu magmatischen Vorgängen. Größere Konzentrationen all der Elemente, die ich in meiner rezenten Klassifikation unter *Ferride* und *Agricolide* zusammengefaßt habe, sind zu finden. Zum anderen sind wertvolle Erzlagerstätten sicherlich nicht auf derzeitige oder ehemalige Orogene beschränkt; große Anteile der weltweiten Produktion an Blei, Zink, Uran, Flour, Barium stammen von den bedeckten Plattformen, zum geringeren Teil Kupfer, Kobalt, Quecksilber und Vanadium. Ein kurzer jedoch unvollkommener Abriß der führenden Provinzen wird weiter unten gegeben, um diesen Punkt hervorzuheben. Spezielle genetische Probleme werden ebenfalls diskutiert.

Die phanerozoische Bedeckung der Plattformen

Die Deckgesteine können in sechs Kategorien unterteilt werden:
1. Karbonate marinen Ursprungs,
2. sandige Ablagerungen, teilweise oder überwiegend marinen Ursprungs; kontinentale Rotsedimente bilden hier die wichtigste Fazies,
3. „coal measures", gewöhnlich mit zahlreichen alternierenden Lagen sandiger toniger Schichten, die Kohlen und Eisenstein enthalten; diese Fazies ist eben-

falls weitgehend non-marin oder brackisch, kann aber u. U. mit Schelfablagerungen verzahnt sein,

4. überwiegend Schiefer, abgelagert unter euxinischen Bedingungen;
5. teilweise Evaporite und verwandte Ablagerungen in abgeschlossenen marinen Becken;
6. Plateau-Laven, überwiegend basaltischen Charakters.

Im Hinblick auf die erzführenden Provinzen sind Fazies 1. und 2. bei weitem wichtiger als alle anderen.

1. Bedeckte Plattformen mit überwiegend karbonatischen Ablagerungen finden sich in dem großen Dreieck, das das Innere des nordamerikanischen Kontinentes bildet. Schwach geneigte oder horizontal gelagerte Dolomite und Kalke vom Kambrium bis Perm bedecken die Fortsetzung des kanadischen Schildes nach Süden. Diese Karbonate sind sicherlich nicht einheitlich mineralisiert. Aber *Heyl* (1967) bemerkt, daß Erzlagerstätten in fast allen paläozoischen Sedimenten auftreten, falls sie nicht erodiert wurden. Es sind jedoch viele Lücken zwischen den zahlreichen Erzdistrikten vorhanden. Alle Haupterzbezirke sind auf Karbonate beschränkt: der Bonneterre-Dolomit (Kambrium) im alten und neuen Bleigürtel Missouris; mittelordovizische Kalke in Wisconsin; Kalke des Mississippian in der Tri State- und Illinois/Kentucky-Provinz; devonische Riffkalke in Pine Point, Kanada. Hinzuzufügen wären die ordovizische Kingsport Formation in Tennessee und der unterordovizische Kalk von Friedensville, wenn, wie viele glauben, die Erzvorkommen der Appalachen in Plattformkalken vor der Orogenese (*Hoagland*, 1971) gebildet wurden.

Die Form der Erzkörper wurde von *Callahan* (1967) zusammenfassend beschrieben. Er zeigte, daß sie a) in einer spezifisch sedimentären Fazies wie z. B. in Talus, auskeilenden Schichten, unterschiedlich kompaktierten Schichten und Riffen auftreten, die mit dem topographischen Relief der unterlagernden Diskordanz in Beziehung steht, b) in Lösungsbrekzien, die zu einer überlagernden Diskordanz in Beziehung stehen (möglicherweise Merkmale einer Verkarstung) und in den charakteristischen Wisconsin-Strukturen auftreten, die durch immer schwächer werdende Lösungen in bestimmten Horizonten gebildet wurden, und c) im Bereich von Fazieswechseln auftreten.

Bemerkenswerterweise sind die meisten Bezirke tektonisch unbeansprucht bis auf einige Ausnahmen, beispielsweise in Illinois/Kentucky, wo Gänge auftreten.

Im großen und ganzen ist die strukturelle Beanspruchung der typischen bedeckten Plattformen im Vergleich zu den Orogenen oder den alten Schilden recht gering. Das Abtauchen in größere Tröge wurde schon erwähnt. Hinzuzufügen wären die gestörten Dome, von denen einige, wie der Ozark Dom, das unterlagernde präkambrische Grundgebirge aufgeschlossen zeigen. *Heyl* (1969) bemerkte, daß die größeren Mineralbezirke mit solchen Vorkommen in Verbindung stehen. Erst kürzlich (1972) untersuchte er die Lineamente im Grundgebirge, die in den überlagernden Gesteinen als ein Verbundsystem von Brüchen und kleineren Strukturen auftreten und über 1300 km in den Mittleren Westen hinein verfolgt werden können. Der Pine Point-Bezirk in Kanada läßt sich ebenfalls mit einem Lineament in Verbindung bringen. Bruchtektonik ist eine strukturelle Eigenart der Plattformen.

In Nordwesteuropa sind die bedeckten Plattformen recht gut in Irland, Nordengland und Belgien zu sehen. Sie setzen sich wahrscheinlich nördlich der variszischen Gebirge in die deutsch-polnische Ebene fort, obwohl sie hier unter mächtigem Pleistozän und anderen nachpaläozoischen Formationen verborgen sind. Weiter im Süden liegen die Blei-Zink-Bezirke Oberschlesiens, die an Kalke des Muschelkalks (Mittlere Trias) gebunden sind und in schwach gefalteten Synklinen der Plattformbedeckung auftreten. Obwohl *Bilibin* (1960) sich sehr dafür einsetzte, die weit verbreitete Mineralisation Rußlands durch eine Abfolge von Orogenesen zu interpretieren, konnte doch *Dimitriyev* (1968) zeigen, daß die Merkmale bedeckter Plattformen über weite Gebiete östlich und westlich des Urals vorhanden sind.

Zusammenfassend kann gesagt werden, daß Karbonate verschiedenen Alters, vom Kambrium bis zur Trias, die die wichtigsten lithologischen Einheiten der bedeckten Plattformen ausmachen, die Provinzen und Bezirke umfassen, die wichtig für Blei-, Zink-, Fluor-, Barium- und in geringerem Ausmaß für die Kupferproduktion sind. Das Auftreten der Erzvorkommen wurde durch Permeabilitätskanäle verschiedener Art kontrolliert, unter welchen Paläogrundwasserleiter (*Hoagland*, 1971) in Verbindung mit Karst sicherlich eine wichtige Rolle spielten. Andere Faktoren sind Klüfte, kleinere Brüche, brekziöse Zonen eines tektonisch bedingten Riffrandes in Verbindung mit Lösungserscheinungen und Diskordanzen (z. B. zwischen Grundgebirge und Bedeckung).

2. Wenden wir uns nun der „*Rotbedeckung*" zu, so geben die Formationen des Unterperm (Rotliegendes), des Buntsandstein und Keuper NW-Europas und der Trias und des Jura auf dem Colorado Plateau ein gutes Beispiel. Obwohl das Colorado Plateau von Elementen des Anden-Orogens umgeben ist, fällt es doch unter die Kategorie der bedeckten Plattformen, wie die allgemein schwach einfallenden, strukturell einförmigen roten Sandsteine andeuten. Dispers auskristallisierte Lagerstätten treten in triadischen Sandsteinen Deutschlands, besonders in Mechernich (Blei) und Commern (Kupfer) auf. Im kleineren Maßstab sind andere Vorkommen aus dem unterpermischen Rotliegenden bekannt geworden. Oberpermische Sandsteine enthalten große Kupfervorkommen, z. B. die vorsudetischen Erzkörper in Polen, die seit dem Krieg entdeckt wurden (*Oberc* & *Serkies* 1968), während weiter im Osten permische Sandsteine Erzvorkommen in Rußland enthalten. Die Alderley Edge-Vorkommen in England führen Kupfer, Vanadium und Uran in Keupersandsteinen. Baryt ist als Zement weitverbreitet in den Sandsteinen Cheshire's und Nottingham's. Um Elgin am Moray Firth in Schottland kommen große unregelmäßige Zonen mit Fluorit sowie etwas Bleiglanz hinzu.

Entsprechende Vorkommen in Rotsedimenten des Colorado Plateaus enthalten Kupfer, Vanadium, Molybdän und sogar etwas Silber und Gold (*Hess* 1933, *Finch* 1933). Besonders wichtig wurde die Gegend durch ihre Uranvorkommen. Die Bedeutung der Vorkommen führte zur intensiveren geologischen Untersuchung, als deren Ergebnis zwei Haupttypen unterschieden werden konnten (*Fischer* 1970): a) die Wyoming Konkretionen, lang ausgezogene Körper, die vertikal ganz oder teilweise Sandsteinabfolgen durchdringen und wie Rollen entlang der Oxydationsgrenze in den Sandsteinen verteilt sind, und b) der fastkonkordante Coloradotyp, geringmächtige, flächenhafte, unstete Körper, die von reduzierten Gesteinen umgeben sind. Ähnliche Erzbezirke treten anscheinend in Rußland auf (*Kashirtseva* & *Sidelnikowa* 1971).

Strukturelle Einflüsse sind im allgemeinen sogar noch geringer in den Rotsedimenten als in den Karbonaten. Der Fluß der mineralisierenden Lösungen folgte in den meisten Fällen Permeabilitätskanälen, die wahrscheinlich während der Diagenese der Sandsteine nicht zementiert wurden. Bruchtektonik und Faltung hat, wie gezeigt werden kann, selten eine bedeutende Rolle gespielt. Die Beziehungen zwischen Bedeckung und Grundgebirge sind in den meisten Rotsedimentbezirken nicht bekannt genug.

3. Paralische Bedingungen, unter denen Kohlen entstehen, brachten geringe syngenetische Bildungen an Sulfiden, ich kenne aber keine wirtschaftlichen Vorkommen dieses Typs. Die zyklischen Ablagerungen des Obervisé und Namur (Karbon) im Norden Englands gleichen den „coal measures", die Flöze sind aber zu dünn, um abgebaut zu werden. Zahlreiche Gänge mit Pb, Zn, F, Ba, treten in dieser Fazies auf, wobei einige die überlagernden „coal measures" des Westfal durchschlagen. Einige Gänge wurden auch im Kohlenbezirk der Ruhr gefunden. Die gut entwickelten echt sedimentären Siderit- und Chamositvorkommen im Mesozoikum Europas gehören einer ähnlichen Fazies an.

4. Schiefer, die in flachen epikontinentalen Becken unter stagnierenden Bodenbedingungen abgelagert wurden, haben Metalle angereichert (*Dunham*, 1961), wirtschaftliche Vorkommen sind aber selten. Das interessanteste phanerozoische Beispiel ist der Kupferschiefer, der in NW-Europa weit verbreitet ist (*Dunham*, 1964). Uran ist in einigen schwarzen Schiefern wie dem schwedischen Kulm angereichert.

5. Evaporitische Bildungen in den Deckgesteinen phanerozoischer Plattformen enthalten keine Metallerzlager; sie sind aber wahrscheinlich bedeutsam als Quellen hypersaliner Minerallösungen, die auf andere Faziesbereiche einwirken. *Davidson* (1966) versuchte, das zu beweisen, aber es besteht keine geographische Verbindung zwischen nichteisenhaltiger oder eisenhaltiger Mineralisierung und den Evaporitbecken.

6. Plateau-Lavagebiete, wie die von NE-Sibirien, des Dekkan, von Antrim-Nord Skye, dem Karroo und Parana sind wichtige Hinweise auf bedeckte Plattformen. Metallogenetisch gesehen sind sie enttäuschend und enthalten selten phanerozoische Sulfidlagerstätten. In einigen Fällen wie in Insizwa erfolgte eine Differentiation der verbundenen intrusiven Decken, die massive Sulfidlagen bildeten; in anderen Fällen (z. B. in Sibirien, *Ivanova*, 1971) ist die Anreicherung von Kupfer erheblich aber noch nicht wirtschaftlich.

Die Deckenbasalte scheinen hauptsächlich aus Spalten ausgeflossen zu sein. Einige zentrale Vorkommen aber, wie der tertiäre Vulkanismus der Hebriden, sind anscheinend an inaktive Kontinentalränder gebunden. Von kleineren Plutonen, die mit ihnen auftreten, abgesehen, fehlen auffallenderweise tiefwurzelnde, größere Intrusionen in den phanerozoischen Plattformen. Das ist bemerkenswert im Falle der Erzprovinzen und -bezirke und fast ohne Ausnahme. Basische und einige ultrabasische Gänge kommen in wenigen Bezirken vor, es ist aber keine genetische Beziehung zum Erz zu erkennen. Das weit verbreitete System basischer Sills im nördlichen England ist sicher nicht die Quelle der hydrothermalen Lösungen in den Penninen *(Dunham,* 1949). Im Mittleren Westen der USA gibt es einige explosive Diatreme. Das Diatrem im Hicks Dome im Illinois/Kenntucky-Bezirk könnte in kausaler Beziehung zur Fluoritmineralisation stehen. Aber die Seltenheit oder gar Abwesenheit solcher Merkmale im großen Blei-Zink-Bezirk läßt vermuten, daß ihnen geringe Bedeutung zukommt. Vielleicht kann dasselbe für die kleinen Alkali-Intrusionen entlang des Lineaments parallel zum 38. Breitegrad gesagt werden. In den bedeckten Plattformen wurden einige wenige Kimberlit- oder Karbonatitdurchschlagsröhren gefunden; einige durchschlagen Gesteine der Kreide in Afrika beispielsweise. Es gibt aber keinen Grund, sie mit der Bildung der Sulfiderze in Verbindung zu bringen. In einigen Karbonatbezirken wurden dünnen, bis sehr dünnen Tufflagen in den Nebengesteinen große Bedeutung beigemessen als Indikatoren einer Beziehung zum Vulkanismus, vielleicht in Verbindung mit vulkanischen Quellen. Solche Lagen scheinen oft weit vom Eruptionspunkt entfernt zu sein, und das Argument ist nicht sehr überzeugend. Was die Deckgesteine anbelangt, so werden ihre Erzvorkommen immer noch am besten von Lindgren beschrieben, der sie nicht in Beziehung zu einer magmatischen Aktivität setzt.

Nichtsdestotrotz müssen wir zugeben, daß wir keine Kenntnis von den Vorgängen im tieferen Teil der Kruste haben. Möglicherweise liegt die Fortsetzung der metamorphen Schilde unter den Deckgesteinen und bildet eine Art Fundament, das örtlich so reaktiviert wurde, daß eine Mineralisierung von außerhalb das einzig sichtbare Zeichen dieser Aktivität in einer alles umfassenden Bedeckung zu sein scheint. Ein Versuch, dieses Problem mittels der Geophysik und Bohrungen in den nördlichen Penninen zu untersuchen, erbrachte eine Granitintrusion in der Mitte der zonaren Abfolge des Blei-Zink-Fluor-Barium-Bezirks *(Dunham* & al. 1965). Der Granit ist allerdings älter als die karbonen Gesteine, die die Vorkommen enthalten. Es ist ebenfalls offenkundig, daß der Wärmefluß, gemessen am Inkohlungsgrad der Kohlen und an Flüssigkeitseinschlüssen epigenetischer Mineralien, stärker als normal ist und noch nicht erklärt werden kann. Es scheint sehr wahrscheinlich, daß in fast allen Erzprovinzen der phanerozoischen Plattformen die Erzausscheidungen bei Temperaturen erfolgte, die 50^0—150^0 C über den in diesen geringen Tiefen normalerweise herrschenden lagen.

Die Vorkommen, die hier zur Diskussion stehen, fallen unter die telethermale Klasse einiger Autoren; nur wenige wären heute kühn genug, sie in der telemagmatischen Kategorie unterzubringen. Von der Mineralogie her sind sie recht einfach: Kupferkies, Pyrit, Markasit, Bleiglanz, Zinkblende, Fluorit, Baryt, Quarz, Kalzit, Ankerit (in Karbonaten); Uraninit, Coffinite, Vanadium-Glimmer, Pyrit, Kupfersulfide, Bleiglanz im Rotsedimenttyp. Die geochemischen Verbindungen sind etwas komplizierter: *Heyl's* (1967) Feststellung für den Mississippi-Typ, daß „Co und Ni vorhanden sind; Ag, As, Sb sporadisch auftreten; Au, Bi, W, Te, Sn, Mo selten sind; Kohlenwasserstoffe weit verbreitet sind", würde ebensogut auf die NW-europäischen Vorkommen zutreffen, wobei Hg in die sporadische Kategorie hinzuzufügen wäre. *Galkiewicz* (1967) berichtet über Cd, Ag, Th, Ge, In, Mo, Hg, As, Sb, Bi, Ni, Mn als Spurenelemente in oberschlesischen Erzen, wobei Kohlenwasserstoffe selten oder gar nicht vorhanden sind.

In vielen, jedoch nicht allen Vorkommen in der Karbonatfazies ist das Bleiisotop im Bleiglanz vom anomalen J-Typ, von dem man annimmt, daß er von radioaktiven Desintegrationsprodukten kontaminiert wurde. Wenn das ein Bindeglied zum Rotsedimenttyp wäre, erscheint es doch seltsam, daß keine bedeutenden Uranvorkommen in der Karbonatfazies gefunden wurden. Es ist ebenfalls bemerkenswert, daß keine großen Zinkvorkommen in der sandigen Fazies auftreten. Wichtig ist, zuerst einmal herauszufinden, in welchem Ausmaß die Sedimentation während der Bedeckung der Plattformen die Entstehung von epizonalen Erzkörpern kontrollierte. Im Falle der oolithischen Eisensteine, die zu mehreren Malen im Phanerozoikum auftraten, gibt es keinen Zweifel, daß das Eisen durch chemische Sedimentation in abgeschlossenen Becken konzentriert wurde, die nur wenig klastisches Material empfingen. Das Eh-Potential war veränderlich, sodaß Chamosit, Siderit und Pyrit gebildet werden konnten, wo reduzierende Bedingungen vorherrschten, Goethit oder gar Magnetit, wo Oxydation vorherrschte. Sphalerit ist weit verbreitet, jedoch unwirtschaftlich in sideritischen Chamositoolithen in Yorkshire, England *(Dunham, 1960)*; andere Metalle, Cu, Pb, Ni, Co eingeschlossen, sind angereichert. Ebenso klar ist, daß ein höherer Gehalt als der Clarke'sche an agricoliden Metallen in einigen Schiefern zu finden ist, wobei die Anreicherung während der Sedimentation von einer Umverteilung während der Diagenese gefolgt wurde, was wohl die beste Erklärung dafür sein dürfte. Viele Autoren (z. B. *Wedepohl*, 1969) glauben, daß Verwitterungslösungen aus exponierten Teilen der Kontinente eine ausreichende Quelle für diese angereicherten Metalle sind. Ich selbst bin überhaupt nicht davon überzeugt *(Dunham, 1964)*, daß der Prozeß erreicht werden kann, ohne daß mineralisiertes Grundwasser daran beteiligt ist, das entweder in Form von Quellen aus dem Seeboden tritt (wie im Roten Meer heutzutage) oder durch Schiefer, die als Membranfilter für solche Grundwässer dienten, nach deren Bedeckung floß. Es wurde der Versuch unternommen, besonders von französischen und deutschen Lagerstättenkundlern

(*Niccolini*, 1970, S. 252), den Mississippi-Typ der Blei-Zink-Fluor-Barium-Vorkommen in der Karbonatfazies auf sedimentären Ursprung zurückzuführen. Ungeachtet der Manifestierung epigenetischer Merkmale, die eigentlich in jedem Vorkommen dieses Typs zu finden sind, hielt man daran fest, daß die Metalle mit den Sedimenten abgelagert wurden. Ich betrachte diese Erklärung als völlig unannehmbar, obwohl in einigen Fällen die Möglichkeit nicht verneint wird, daß Mineralquellen am Seeboden die Erzentstehung in der Karbonatfazies gefördert haben. In der großen Mehrzahl solcher Vorkommen wurde der Gehalt durch Permeabilitätskanäle von außen in das Gestein gebracht. Das Ausmaß echter Lateralsekretion — z. B. Konzentration verstreuter seltener Elemente in den Gesteinen — ist immer noch recht unklar.

Der größte Fortschritt in den letzten zwei Dekaden wurde in den Bereichen der Chemie, der Temperatur und dem Druck von Lösungen gemacht, die für die Konzentration verantwortlich sind. Die Untersuchungstechnik an Flüssigkeitseinschlüssen, zuerst vom englischen Petrographen *Sorby* 1855 angewandt, aber seither lange vernachlässigt, hat wieder durch *Ermakov* in Rußland, *Newhouse*, *Roedder* und andere in den USA an Bedeutung gewonnen. Bildungstemperaturen sind nun hinreichend bekannt und liegen im Bereich von 50⁰—150⁰ C mit einigen Beispielen in der Karbonatfazies, die 200⁰ C überschreiten, ungeachtet der geologischen Aussage, daß die Ablagerung in Tiefen von 1 km oder weniger stattfand. Gefrierexperimente weisen auf hohe Salinität hin, und die hervorragende Arbeit von *Hall* & *Friedman* (1963) an Illinois' Mineralien zeigt, daß die Minerallösungen aus hypersalinen Mineralquellen stammen, die sich gut mit Formationswässern vergleichen lassen, die gewöhnlich unterhalb 700 m in der Erdölexploration (*White* 1968; *Dunham* 1970) gefunden werden. Solche Wässer sind ausgezeichnete Löser für Metalle in Form von Cloridkomplexen.

Obwohl magmatische Aktivität zweifelsohne residuale hypersaline Mineralquellen hervorbringen kann, in denen fast alle löslichen Komponenten des Magma konzentriert sein können, und obwohl, wie *Holland* (1972) zeigte, Mineralquellen einen großen Anteil eines Metalls wie z. B. Zink aus einer magmatischen Abfolge extrahieren können, ist es doch nicht möglich, hydrothermale Lösungen, die von einem juvenilen Magma stammen, anzuführen, um die weit verbreiteten Erzprovinzen in phanerozoischen Plattformen zu erklären. Ein Beitrag von dieser Seite kann jedoch nach *White* (1968) nicht ganz ausgeschlossen werden, aber die Vermutung von *Heyl* (1967) und mir (1967), daß eine Anreicherung an K in Mineralquellen, gemessen an Flüssigkeitseinschlüssen, beweise, daß solche Beiträge stattgefunden haben, ist nicht vereinbar mit den Gesetzmäßigkeiten der physikalischen Chemie (*Helgeson* in einem Diskussionsbeitrag zu *Heyl* 1967). Auch die H/D-Verhältnisse tragen nicht zu einer Lösung des Problems bei. Der größere Teil der Wässer scheinen entweder connate (fossiles Meerwasser) oder atmosphärische Wässer zu sein, die eine extensive Membranfiltration bei erhöhten Temperaturen durchliefen oder eine Kombination von beiden sind.

Die genaue Herkunft der Metalle konnte bisher noch nicht demonstriert werden, und mehr geochemische Forschung an sedimentären Systemen ist notwendig. Karbonate sind nicht von sich aus gute Sammler agricolider Elemente (siehe z. B. *Wedepohl* in *Lavery & Barnes* 1971), obwohl man vermutet hat, daß einige Korallen Zinkanreicherung begünstigen können. Schwarze Schiefer sind vielversprechender als intermediäre Quelle, nimmt man an, daß sie von zirkulierenden Mineralwässern ausgelaugt wurden; aber auch das ist immer noch fraglich. Sandsteine scheinen wegen ihrer großen Permeabilität vielversprechender zu sein. Vor kurzem führten *Doe & Delevaux* (1972) aufgrund ihrer Untersuchungen an der isotopischen Zusammensetzung von Blei an, daß der große Missouri Bleigürtel im Bonneterre-Dolomit sein Blei von Mineralwässern bekommen haben könnte, die den unterlagernden Lamotte-Sandstein auslaugten. Weitere Versuche, Isotope als Spurensucher in dieser Art anzuwenden, wären sehr wünschenswert.

Schwefel wird nicht unbedingt von den gleichen Mineralwässern herangebracht, obwohl es der Fall sein könnte. *Gerdemann & Myers* (1972) behaupten, daß der Schwefel in SE-Missouri aus der Dekomposition von Algenriffen oder anderen organischen Sedimenten herstamme. *Skinner* (1967) nahm schon früher an, daß der Schwefel von organischem Material und Erdöl stamme, wobei heiße Mineralquellen in den Mississippi-Valley-Bezirk eindrangen.

White (z. B. 1968) führte wiederholt an, daß das Mischen von Mineralwässern verschiedenen Ursprungs eine bedeutende Rolle in der geochemischen Evolution der Provinzen spielte. Modelle, die auf dieser Grundlage basieren, um die Mineralisation der englischen Penninen zu erklären, wurden von *Sawkins* (1966) und *Solomon, Rafter & Dunham* (1972) vorgebracht, wobei letzterer sich auf Isotopendaten stützte.

Man muß sich fragen, wie die Mineralwässer, seien sie connate oder atmosphärisch-gefiltert, aufgeheizt werden. Einige Autoren nehmen einfach nur Perioden höherer geothermaler Gradienten zur Hilfe, aber man kann sich das ohne magmatische Aktivität in der Tiefe kaum vorstellen. Die andere Möglichkeit wäre ein Absinken der Mineralwässer aufgrund ihrer Schwere zu einer Tiefe von zehn oder mehr Kilometern, bis sie durch Tektogenese zurückgetrieben werden; oder das Vorhandensein connater Wässer, die erst tief versenkt, und wenn Beanspruchungen auftreten, ausgetrieben werden.

Wenden wir uns abschließend den Uran-Vanadium-Kupfer und Blei-führenden Sandsteinen zu, so nahm man für die ersteren sauerstoffreiche Wässer als Medium an; das gilt für die Colorado Plateau Vorkommen, die frühere Autoren als sedimentär *(Warren 1972)* betrachtet hatten. Man kann die gleiche Erklärung nicht auf alle Sandsteine anwenden. Interessanterweise aber widersprach *Nuralin* (1964) den sedimentären Hypothesen *Strakhovs* (1960) und *Konstantinovs* (1963), daß die russischen Kupfersandsteine syngenetisch und sedimentär seien. Eine Konferenz über dieses Problem in Moskau 1965 fand diese Hypothesen ebenfalls unannehmbar. *Bezrodnyikh* (1971) führte aus, daß die Erzkonzentrationen im südlichen Teil der sibirischen Plattform postdiagenetisch entstanden.

Schlußfolgerungen

Die Erzprovinzen und -bezirke, die man charakteristischerweise in den phanerozoischen bedeckten Plattformen findet, sind mineralogisch einfach. Geochemisch beinhalten sie gewisse Elemente, die bemerkenswerterweise in nur geringen Konzentrationen auftreten, wie Cr, Sn, W, Ni (abgesehen von Plateaulaven), B. Von vielen dieser Vorkommen nimmt man an, daß sie durch warme oder heiße hypersaline Mineralwässer, die über erhebliche Tiefen in der Kruste zirkulierten, abgelagert wurden. Die Elemente, die nun in konzentrierter Form auftreten, wurden aus Sedimenten (die bevorzugte Abfolge scheint die folgende zu sein: 1. Sandsteine, 2. schwarze Schiefer, 3. Karbonate) oder möglicherweise aus magmatischen oder metamorphen Gesteinen des unterlagernden Grundgebirges ausgelaugt. Die Ränder großer Sedimentbecken — wie in Illinois, Michigan, der Nordsee, NW-Rußland — sind vielleicht die bevorzugtesten Gebiete für solche Provinzen.

Literatur

Bezrodnyikh, Y. P. (1971): The Process of formation of cupreous sandstones in the light of geochemical data. — Abs. Int. Geochem. Cong., Moscow, 2, 884—5.
Bilibin, J. A. (1960): Die geochemischen Typen den orogeren Zonen. — Z. f. angew. Geol., *6,* 545—49.
Blondel, F. (1937): La geologie et les mines des vieilles platformes. — Paris.
Callahan, W. H. (1967): Some spatial and temporal aspects of the localisation of Mississippi Valley-Appalachian type ore deposits. — Econ. Geol. Monog. *3,* 14—19.
Davidson, C. F. (1966): Some genetic relationships between ore deposits and evaporites. — Trans. Inst. Min. Metall., *75,* B215—25.
Dmitriyev, L. M. (1968): Polymetal mineralization in northeastern Russian Platform and adjacent mobile folded zones. — Internat. Geology Rev., *10,* 4.
Doe, B. R. & Delefaux, M. H. (1972): Source of lead in Southeast Missouri galena ores. — Econ. Geol. *67,* 409—25.
Dunham, K. C. (1949): The Geology of the Northern Pennine Ore Field. Vol. I. Tyne to Stainmore. — Mem. Geol. Survey. 357 pp, 4 plts, 33 figs.
Dunham, K. C. (1960): Syngenetic and diagenetic mineralization in Yorkshire. — Proc. Yorks. Geol. Soc., *32,* 229—284.
Dunham, K. C. (1961): Black Shale, Oil and Sulphide Ore. (Presidential Adress delivered 1. 9. 1961). — Advancement of Science, *18,* No. 73, 274—99.
Dunham, K. C. (1964): Neptunist concepts in ore genesis. — Econ. Geol. *69,* 1.—21.
Dunham, K. C., A. C. *Dunham,* B. L. Hodge & G. A. L. *Johnson* (1965): Granite beneath Viséan sediments with mineralization at Rookhope, northern Pennines. — Quart. J. Geol. Soc. London, *121,* 383—417.
Dunham, K. C. (1970): Mineralization by deep formation waters. — Trans. Instn. Mining Metall. (Sect. B Appl. earth sci.) *79,* Bl 27—36.
Dunham, K. C. (1972): Basic and applied geochemists in search of ore. — Ninth Sir Julius Wernher Memorial Lecture. Trans. Inst. Mining and Metal.
Dunham, K. C. (1973): Geological Controls of Metalliferous Provinces. — 12th Pacific Sci. Congress, Symposium D, Bur. Min. Res. Australia, Bull. *141,* 1—11.
Finch, J. W. (1933): Copper deposits of the Western States. — *Lindgren* Volume: Ore deposits of the Western States, 481—87.
Fischer, R. P. (1970): Similarities, differences and some genetic problems of the Wyoming and Colorado Plateau types of uranium deposits in sandstones. — Econ. Geold. *65,* 778—85.

Galkiewicz, T. (1967): Genesis of Silesian-Cracovian zinclead deposits. — Econ. Geol. Monog. *3*, 156—77.

Gerdemann, P. E. & *Miers*, H. E. (1972): Relationship of carbonate facies patterns to ore deposition and to ore genesis in the south-east Missouri lead district. — Econ. Geol., *67*, 426—33.

Guild, P. W. (1971): Metallogeny: a key to exploration. — Min. Engng. N.-Y., *23*, 69—72.

Hall, W. & *Friedman*, I. (1963): Composition of fluid inclusions, Cave-in-Rock district, Illinois and Upper Mississippi Valley zinc-lead district. — Econ. Geol., *58*, 886—911.

Hess, F. L. (1933): Uranium, vanadium, radium, gold, silver and molybdenum sedimentary deposits. — *Lindgren* Volume: Ore Deposits of the Western States, 450—81.

Heyl, A. V. (1967): Genesis of stratiform zinc-lead-barite-fluorite deposits. — Econ. Geol. Monog. *3*, 20—32.

Heyl, A. V. (1969): Some aspects of genesis of zinc-lead-barite-fluorite deposits in the Mississippi Valley, USA. — Trans. Inst. Min. Metall., *78*, B 48—60.

Heyl, A. V. (1972): The 38th Parallel lineament and its relationship to ore deposits. — Econ. Geol. *67*, 879—94.

Hoagland, A. D. (1971): Appalachian strata-bound deposits: their essential features, genesis and the exploration problem. — Econ. Geol. *66*, 805—10.

Holland, H. D. (1972): Granites, solutions, and base metal deposits. — Econ Geol. *67*, 281—301.

Ivanova, A. (1971): Content and distribution of trace elements in traps of the northwestern Siberian platform. — Abst. Int. Geochem. Congr., Moscow, *1*, p. 144.

Kashirtseva, M. F. & V. O. *Sidelnikova* (1971): Selenium, uranium, molybdenum in oxogenous roll-type orebodies. — Abst. Int. Geochem. Cong. Moscow, *2*, 890—1.

Kelly, W. C. & *Turneaure*, F. S. (1970): Mineralogy, paragenesis and geothermometry of the tin and tungsten deposits of the Eastern Andes, Bolivia. — Econ. Geol. *65*, 709—80.

Konstantinov, M. M. (1963): Proiskozhdeme Stratifitsirovannykh Mestorozhdenii Svintsa 1. Trinka. — Academy of Sciences, Moscow, 183 pp.

Mitchell, A. H. G. & M. S. *Garson* (1972): Relationship of porphyry copper and circum-Pacific tin deposits to palaeo-Benioff zones. — Trans. Inst. Min. Metall., *81*, B 10—25.

Niccolini, P. (1970): Gitologie des concentrations minerales stratiformes. — Paris, 792 pp.

Nuralin, N. N. & al. (1964): Geologiya Rudnikh Mestorozhdenii *6*, 105—12.

Oberc, J. & *Serkies*, J. (1968): Evolution of the Forc-Sudetian copper deposit. — Econ. Geol. *63*, 372—79.

Petrascheck, W. E. (1965): Typical features of metallogenic provinces. — Econ. Geol., *60*, 1620—34.

Sillitoe, R. H. (1972): A plate tectonic model for the origin of the porphyry copper deposits. — Econ. Geol., *67*, 184—97.

Skinner, B. J. (1967): Precipitation of Mississippi Valley type ores: a possible mechanism. — Econ. Geol. Monog. *3*, 363—370.

Solomon, M., T. A. *Rafter* & K. C. *Dunham* (1971): Sulphur and oxygen isotope studies in the northern Pennines in relation to ore genssis. — Trans. Inst. Mining Metall. B, *80*, 259—76.

Strakhov, N. M. (1960): Osnovy Teorii Litogeneza, *1*, Moscow.

Warren, C. G. (1972): Sulfur isotopes as a clue to the genetic geochemistry of a roll-type uranium deposit. — Econ. Geol., *67*, 759—67.

Wedepohl, K. H. (1969) *ed.:* Handbook of geochemistry. Berlin, *1*, 442 pp.

Wedepohl, K. H., N. G. *Lavery* & H. L. *Barnes* (1971): Zinc dispersion in the Wisconsin zonc-lead district. — Econ. Geol., *66*, 226—242.

White, D. E. (1968): Environments of generation of some base-metal deposits. — Econ. Geol., *63*, 301—35.

Palaeomagnetism and Ore Deposits

D. H. *Tarling*[*]

Abstract

The ways by which rocks acquire a magnetisation can be used to distinguish the thermochemical history of ore deposits and offers a tool for both relative and absolute dating. In addition, "standard" palaeomagnetic techniques can be used to determine the structural development of the area in relationship to neighbouring regions, as well as assisting in determining the internal evolution of an ore deposit.

Zusammenfassung

„Paläomagnetismus und Erzlagerstätten"

Das Vorkommen von Erzlagerstätten in Zusammenhang mit Bewegungen von Kontinenten (oder Platten) muß von zwei Gesichtspunkten betrachtet werden. Der einfachste Fall ist dort gegeben, wo bestehende Lagerstättenprovinzen aufgespalten und die Teile in verschiedenen Kontinenten separiert werden wie z. B. die Nickellagerstätten von Australien — Antarktis — Indien — Afrika oder die Diamantenfelder von Westafrika und Guyana. Der kompliziertere, aber häufigere Fall ist der, daß beim Aufbrechen der Kontinente Bedingungen geschaffen werden, unter welchen Erze infolge eindringender magmatischer Lösungen entlang der Plattenränder angereichert werden. Beispiele hiefür sind die Blei-Zink-Silber-Lagerstätten von Irland und Norwegen oder das Rote Meer. Paläomagnetische Untersuchungen sind hier, indem sie die ursprüngliche Lage der Kontinente bestimmen, von entscheidender Bedeutung. Das gilt vor allem auch für Lagerstätten, die weitgehend von den palöoklimatischen Bedingungen abhängig sind, wie z. B. Bauxit-, Gips- und möglicherweise auch Erdöl- und Erdgas-Lagerstätten.

[*] D. H. *Tarling*, Dept. Geophysics & Planetary Physics, University of Newcastle upon Tyne, NE1 7RU, England.

Für die letzten 75 Millionen Jahre sind die paläomagnetischen Messungen des Ozeanbodens die genaueste Methode, um frühere Lagen der Kontinente zueinander zu erfassen. Diese Methode kann höchstens bis in den Jura oder vielleicht bis in die Trias ausgedehnt werden, wenn eine genauere Datierung der inversen Folge im Mesozoikum möglich ist. Für die Zeit des Mesozoikums und jedenfalls für die älteren Zeitepochen sind die paläomagnetischen Messungen an kontinentalen Gesteinen für die Bestimmung der Lagen der Kontinente und somit auch für die Festlegung des Beginns des Auseinanderbrechens von Kontinenten oder des Zeitpunktes von Kollisionen zwischen Teilen von Kontinenten von entscheidender Bedeutung.

Die Genauigkeit solcher Rekonstruktionen ist zur Zeit noch viel stärker von der Verläßlichkeit der Ausgangsdaten als von den theoretischen Grundlagen der paläomagnetischen Methode abhängig. Dennoch besteht die Möglichkeit, Hoffnungsgebiete für Erzlagerstätten anzuzeigen.

1. Introduction

Palaeomagnetic studies of continental and oceanic rocks (*Irving*, 1964; *Tarling*, 1971; *McElhinny*, 1972) have played a crucial role in recent changes in the concept of continental drift, and clearly palaeogeographic reconstructions based on these concepts are likely to be of some value in relating ore deposits within and between continents. More significantly, ore minerals may either consist of ferromagnetic minerals, such as heamatite and magnetite, or contain them in significant quantities as accessory minerals. Thus a study of the origin of the magnetisation of these minerals can, in many cases, be used to investigate the physical condition under which the ores were deposited and, if geomagnetic field changes are resonably well documented for other rocks in the region, it should be possible to obtain relative or absolute dates for the ore deposits by direct comparison of their direction of remanence with those of rocks of known ages. Similarly, the application of palaeomagnetic techniques may yield information on the gross tectonic evolution of the ore field or assist in petrofabric analyses where standard techniques appear to be less effective.

This article is mainly concerned with the way in which rocks acquire their magnetisation (Section 2) and how this can be used for the dating of ore bodies (Section 3). Palaeolatitude, tectonic and petrofabric considerations are only considered briefly in section 4.

2. The Acquisition of Remanent Magnetisation

The ways in which rocks acquire a remanent magnetisation are now fairly well known although it may be difficult to assess the major factors involved in any one particular example, particularly in the case of ore deposits. Essentially two main physical processes are involved. As an igneous rock cools from its molten state, it solidifies and eventually passes through the Curie temperature of its component magnetic mineral. This temperature, mostly below 680⁰ C in rocks, is where the quantum mechanical forces between iron atoms become sufficient to couple their electron spins so that they are brought into alignment with any external field, despite the disrupting action of thermal vibrations. The length of time during which this alignment is maintained is the *relaxation time;* the time it takes for the direction of remanent magnetisation to change from its original direction to that of a newly applied field, or to become randomly oriented in the absence of an external field. The relaxation time (τ) of a magnetic grain is dependent on its composition (K) and also on its volume (V) and temperature (T) — $\tau \, \alpha \, K \, (V/T)^{1/2}$ — i. e. below its Curie temperature its relaxation time increases approximately exponentially for a linear increase in volume or decrease in temperature (Néel 1955). An increase in volume or decrease in temperature can therefore result in the preservation of a direction of remanence throughout geological time; the temperature or volume at which the grain acquires a relaxation time of the order of a few minutes is termed the "blocking" temperature or "blocking" volume i. e. when the relaxation time corresponds to the duration of normal laboratory experiments. Large particles (haematite greater than 10^{-3} cm diameter, titanomagnetite greater than abouth 10^{-5} cm diameter) also develop strong magnetostatic forces because of the separation of magnetic poles on opposite ends of the grain. These cause the atomic alignments within the grain to form antiparallel groupings (domains) and such multidomain particles can change their net direction of magnetisation by unrolling the "wall" between each domain, so such large particles are generally less magnetically stable than single domain particles. As atoms within grains are able to migrate, especially at high temperatures ($> 600^0$ C), this allows the exsolution of the individual large titanomagnetite grains into lamellae of, for example, ulvospinel and magnetite or ilmenite and haematite. The individual magnetite or haematite exsolutions may be of single domain size, physically isolated from each other by a non-magnetic matrix and thus magnetically very stable, although the original particle was large.

As an igneous rock cools, a stable remanence is acquired by the single domain particles, in particular, so that when these particles are eroded and subsequently deposited as a wet slurry, they become physically aligned along the Earth's mangetic field direction. Subsequent cementation may preserve this direction, but the diagenesis is also associated with major chemical changes in magnetic minerals, particularly the dehydration of clay minerals, goethite, etc., to form haematite, and the oxidation of other iron-bearing minerals (olivines, pyroxenes, amphiboles, etc.) As the haematite or magnetite grows, individual grains acquire a remanence directed along the ambient magnetic field as they pass through the blocking volume and, if growth continues, this direction may be preserved indefinitely. Sedimentary rocks are therefore likely to carry a remanence associated

43

with both deposition and diagenesis, so that the age of the stable remanence is normally only that of the age of formation the rock if diagenesis took place shortly after deposition.

Many rocks may therefore contain ferromagnetic particles which carry a remanence acquired when the rock originally formed, but the wide range of grain sizes usually means that a wide spectrum of relaxation times is present. Grains of low relaxation time will have lost their original direction, but will also have acquired a magnetisation in later magnetic fields. However, the magnetisation of such short relaxation time grains can be preferentially removed if rock samples are subjected to partial demagnetisation by incremental temperatures or alternating magnetic fields. Incremental demagnetisation can therefore isolate the remanence associated with grains of long relaxation times in many rocks. However, this critical size range may be absent or consist of such a small component of the total magnetic fraction that instrumental defects (such as a few gammas direct field while cooling during thermal demagnetisation or even harmonies in the current producing the alternating magnetic field) may mask any high stability component. This has meant that most standard palaeomagnetic work has hitherto been mainly concerned with basic to intermediate igneous rocks and red siltstones, but technical improvements in both measurement and demagnetisation is allowing a wider range of rock types to be used in palaeomagnetic studies. Comparatively few studies have so far been undertaken on either metamorphic rocks or ore bodies (although some particular examples are mentioned below). This arises because the origin of remanence in such rocks is likely to be complex and also the direction of remanence which was acquired may not exactly parallel the ambient field because the magnetic particles carrying the remanence are likely to be inhomogeneously distributed within the rock samples and there may be a net alignment of either the crystallographic axes or grain shapes which causes the direction of remanence to be pulled towards an "easy" directions of magnetisation, i. e. along the [111] axis of aligned magnetites or the net major axis of elongate particles. These anisotropic effects can be used in magnetic structural analyses (Section 4). An elongation of macroscopic or microscopic grains may not necessarily apply to the submicroscopic domain size particles which carry the most stable remanence and there is evidence that anisotropic effects are reduced by partial demagnetisation. Nonetheless in the initial development of palaeomagnetic studies, samples of rocks in which the anisotropy was in excess of 5 % (max. susceptibility/min. susceptibility greater than 1.05) were generally ommitted from further study (*Irving* 1964).

In the case of metamorphic rocks and ore deposits, heamatite and magnetite are common accessory minerals but the processes of magnetisation must involve varying degrees of thermal heating and crystal growth, often while the rocks are under pressure. The main effect of pressure appears to be the creation of directions of preferred crystal growth which enhance anisotropic properties, but direct magnetic effects are mostly lost when the pressure in removed. The slow cooling and prolonged crystal growth in many ore bodies means that the total remanence for a single ore sample may have been acquired over a prolonged period and therefore reflects the average of possibly complex geomagnetic field changes. Nonetheless, it is becoming increasingly clear that the palaeomagnetic study of ore minerals, and their associated rocks, should yield information on the processes and rate of ore accumulation.

3. Palaeomagnetic Dating

As magnetically isotropic rocks acquire a magnetisation parallel to the geomagnetic field at the time they pass through their blocking temperature or blocking volume, it is possible to use changes in the direction, and, to a lesser extent, the strength of the geomagnetic field as a means of dating this event. The Earth's magnetic field shows an extremely wide spectrum of changes, from microsecond pulsations to polarity reversals and polar shifts over 10^7 years and greater. In general most metallic ore bodies take at least a 1000 years to form so that all changes of less than 1000 year periodicity are likely to be averaged out. Therefore dating by comparison with short term secular changes, which is applicable in archaeological studies, is not likely to be relevant, except under exceptional circumstances. Studies of such changes, however, indicate that on longer time scales, these secular changes average out to yield an average geomagnetic field which is very close to that of a single axial geocentric dipole and this concept (discussed further in Section 4) is assumed to be valid for most of geological time. This means that it is possible to allow for the spatial variation in the geomagnetic field by defining the palaeomagnetic direction in terms of the corresponding pole position. (This is, in fact, merely another mathematical way of describing the direction of remanence and it is a different philosophical step to attribute this pole position to an actual geomagnetic pole. Over a small region, a few hundred km², the spatial variation of the geomagnetic field is small and palaeomagnetic directions can be compared directly.) With increasing geological time, the average pole positions for each stable tectonic block become increasingly displaced from the present rotational pole, forming a polar wandering curve (Figure 1), but each tectonic block has its own distinct curve so that, for example, the polar wandering curves of the last 300 million years for Europe and North America, derived from rocks unaffected by orogenic activity, follow a similar pattern, but only begin to converge during the last 180 million years. Many palaeo-continental reconstructions, of course, depend on matching such polar wandering curves. When such curves are available, it is possible to obtain average directions for rocks of unknown age on the same block and, by comparison with the established polar wandering curve, the time at which they acquired their remanence can be determined — an example in relation to ore bodies is given by *Hanuš & Krš* (1963). Where such curves have not been established, it is still

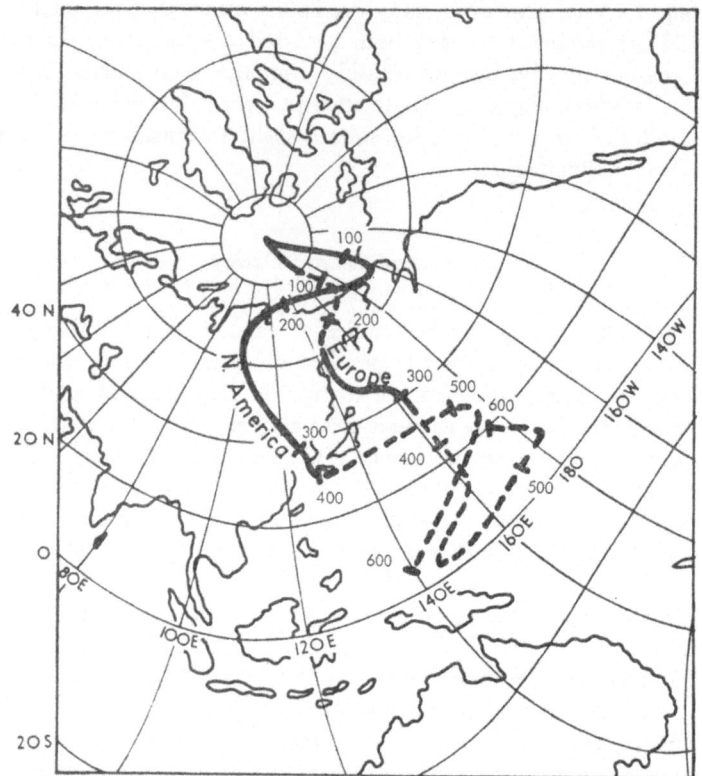

Fig. 1. Polar Wandering Curves of Europe and North America. The solid lines indicate well documented pole determinations, and broken lines indicate where there is still uncertainty about the precise pole positions.

possible to obtain the relative ages of different rocks by comparison of their directions. The injection of dykes in some parts of the Canadian Shield, for example, appears to have been approximately contemporaneous as most of the dykes have a very similar direction of remanence, so that there cannot have been much movement between the shield and pole during the time the dykes were injected *(Larochelle, 1967)*. Conversely studies of haematite bodies in the Lake Superior region have shown that some ore bodies have directions similar to intrusive rocks and are therefore probably syngenetic, while others have significantly different directions and are therefore probably residual weathering products *(Symons, 1967 a, b)*.

The accuracy of this method of absolute dating is very difficult to evaluate as precise polar wandering curves have not yet been satisfactory established for most continents, and the precision will also depend on the rate at which the sampled area and pole are moving relative to each other. Nonetheless, it seems likely that an accuracy of some 10^7 years can be achieved by this method if at least 50 separately oriented samples with stable remanence, can be collected for a time of relative polar movement of some 0.5º/m. y. (the average relative polar movement is 0.3º/m. y. for Europe during the Mesozoic-Cenozoic).

More precise dating is possible using reversals of the polarity of the geomagnetic field. These were first identified at the start of this century, but detailed study of the age sequence of polarity changes has only been achieved during the last decade. These have shown that the reversals take place irregularly, although possibly statistically conforming to a Markov chain sequence *(Parks, 1972).* The events during an actual transition of polarity are only poorly known, but it appears that there is a decrease in intensity of the geomagnetic field over a period of some 4000 years, when it falls to approximately on fifth of its usual value. The actual polarity change then takes place over an interval, possibly as short as 2000 years, and is followed by a gradual return to its usual intensity over another 4000 years. This polarity is then retained for 10^4 to 10^7 years. The polarity transition is, therefore, geologically instantaneous and world-wide and forms an ideal dating horizon. Further study may show that the precise nature of each reversal may be sufficiently different that detailed studies may allow their specific identification, but this is unlikely to be applicable for some decades, if at all, and is probably of too short a duration to be isolated during the formation of ore bodies. Nonetheless it is possible to match polarity sequences or, if the approximate age is already known, the presence of reversals may allow more precise dating than either normal palaeontological or radiometric methods. One particular example is the magnetisation acquired by deep-sea manganese nodules which have recently been shown to contain both normal and reverse polarities *(Crecelius, Carpenter & Merrill, 1973).* The normal polarity was acquired during the last 700,000 years, i. e. during the present normal polarity period, and the reversed components are older, confirming the very slow rate of growth of these mineral deposits. Similar dating, with very high precision, is possible through most of the Cenozoic as the polarity sequence is well documented from oceanic magnetic anomaly patterns *(Heirtzler & al., 1968; Tarling, 1971)* and fuller information is gradually becoming available for much of the late Palaeozoic and Mesozoic (Figure 2). It is known that the Earth's magnetic field was almost entirely of reversed polarity in the late Carboniferous and for most of the Permian (280 to 230 m. y.) but was equally normal and reverse during most of the Triassic (190—230 m. y.), and mostly normal during the Jurassic and Cretaceous. Thus it is possible to distinguish between Permian and Triassic sandstones, or ore bodies, using the ratio of normal to reversed rocks as long as a several million year interval is sampled.

It is known that certain minerals of specific composition have the ability to self-reverse, i. e. on cooling they acquire a direction of remanence in opposition to that of the ambient field. Similarly, intra-crystalline re-arrangements over

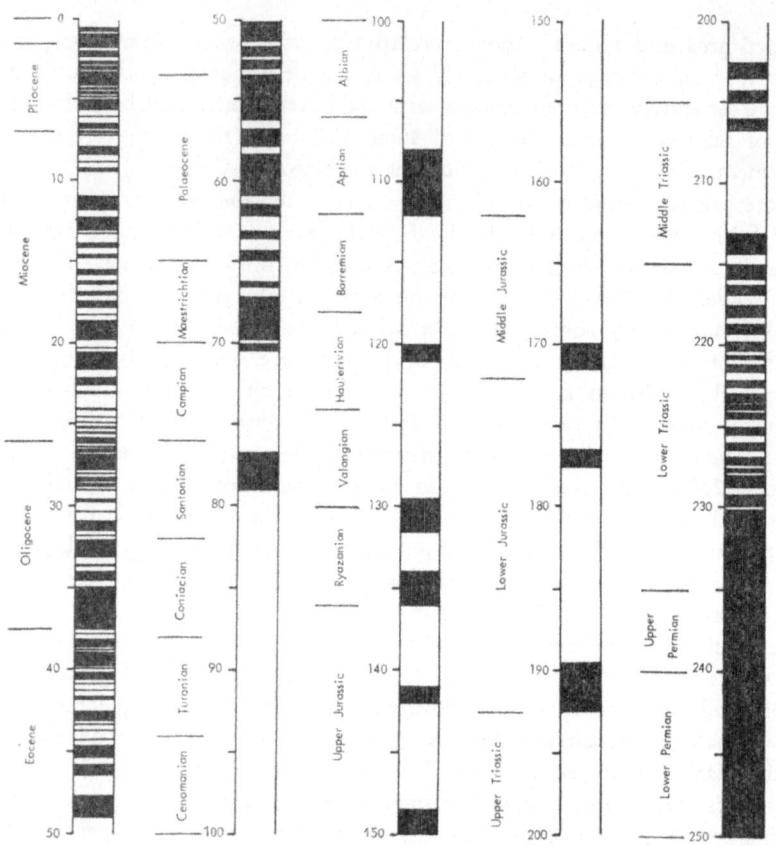

Fig. 2. The Mesozoic-Cenozoic Polarity Sequence. The white bands correspond to present day "normal" polarity and the black correspond to "reversed" polarity zones. The Cenozoic scale is estimated to be accurate within some 3 %, but the detailed timing of Mesozoic polarity changes is subject to greater modifications as further polarity changes are documented, but mainly as dating techniques and data are improved.

geological time may cause the magnetic polarity of the sample to reverse. However, these processes are now fairly well understood *(Ishikawa & Syono, 1963)* and are restricted to rare and specific mineral compositions under abnormal physical conditions. Thus it is extremely unlikely that such self-reversals will be encountered and only one well authenticated naturally self-reversing rock is known, the Haruna dacite of Japan *(Nagata, Uyeda & Akimoto, 1952)* despite the many thousands of rocks which have been examined for this property.

4. Geotectonic and Palaeolatitude Applications

Rocks which acquired their remanence at the same time, have not moved relative to each other subsequently, and carry a stable remanence, must have directions of magnetisation which reflect the original geomagnetic field. Studies of archaeological material, lake and glacial sediments deposited during the last 10,000 years, deep-sea sediments magnetised during the last 3 m. y., lavas and dykes formed during the last 20 m. y. all show individual directions which correspond to pole positions scattered around, but strongly centred upon the Earth's present axis of rotation. These numerous observations therefore confirm theoretical considerations for an axial geocentric model of the average geomagnetic field. However, individual studies of igneous rocks have also shown that during some of this time, the Earh's average field may be more closely described by an axial dipole that is displaced northwards by some 200—300 km from the Earth's centre, thereby causing the poles of individual collections to consistently lie some 4^0 on the far side of the pole (Wilson, 1970, 1971). This raises some problems about the precise nature of the geomagnetic field and whether this effect is in fact a reflection of other tectonic processes rather than invalidating the axial geocentric dipole model. Nonetheless, for most palaeogeographic reconstructions, this discrepancy is small and is within the statistical limits of almost all average pole determinations. For earlier times, the movement of different tectonic blocks during continental drift, etc., means that global comparisons cannot be made using the present distribution of the continents, but the available palaeomagnetic data remains consistent with a geocentric dipole model on a continental scale. An agreement with such generally accepted palaeoclimatic indicators as coral reefs, archaeocyathids, dolomites, red beds, evaporites, etc., suggests that this dipole is also axial. Unfortunately these palaeoclimatic indicators can only be used as a rough guide as they are sensitive to local geographical factors, global varations in climate, etc. (Conversely, palaeoclimatic conditions should not be inferred precisely from precise palaeomagnetic determinations of palaeolatitude determined using the formula, tan (palaeolatitude $=$ $^1/_2$ than [inclination of remanence]). Nonetheless, it is clear that the axial geocentric dipole model for the average geomagnetic field is a very close first approximation for most of geological time, with the noteable, but brief exceptions of intervals of polarity transitions. It is therefore possible to make first order geological interpretations of palaeomagnetic observations in this way.

In terms of ore deposits, this means that approaching continents must be separated by an ocean which includes an active subduction zone with its associated ore deposits (Guild, this issue; Mitchell & Garson 1972; Sawkins, 1972). Past subduction and collison zones can therefore be documented from palaeomagnetic studies. Similarly, parts of oceanic crust and mantle may become incorporated into continental rocks during continental closures, affording potential ore sources from already differentiated and deposited minerals, such as the copper ores of

Cyprus, Turkey, etc., as well as rocks which, when subjected to chemical weathering, etc., may develop localised enrichments, such as the nickel and chrome deposits of New Caledonia. Ores may also be transported into stable continental areas in which palaeoclimatic considerations may be of major significance in creating optimum environments for their deposition and concentration *(Tarling* 1973). Naturally, similar palaeoclimatic considerations apply even more directly where deposits of evaporites, phosphates, etc. are involved. These factors are clearly more significant than simple jig-saw considerations of continental drift, such as the prediction of ore deposits in Antarctica from those mapped in Australia, or gold and diamonds in Africa matching those in South America etc.

On a more localised scale, the same techniques can be applied to unravel the tectonic developments within individual continental blocks, so that, for example, the relative movements between extra Alpine Europe and the Iberian peninsula, Corsica-Sardinia, etc., are becoming better documented following palaeomagnetic studies and these "local" tectonic events have a clear significance in understanding, for example, the origin for the mineralisation in Sardinia and Tuscany. Local tectonic interpretations, based on palaeomagnetic studies, have also been made to determine, for example, if the Sudbury nickel deposits were laid down in approximately their present position, or were originally quasi-horizontal and have subsequently been folded *(Hood* 1961), and there are numerous similar structural problems to which palaeomagnetic techniques may afford both direct and indirect information. As both haematite and magnetite are stress sensitive minerals, it is also possible that current studies of the detailed magnetic structure of rock samples may quickly and simply yield petrofabric information relevant to past stress fields which is not discernable from standard petrofabric analyses of the orientation quartz grains, micas, etc.

5. Summary

An increased understanding of the processes by which rocks acquire their magnetisation now means that the study of the remanent magnetisation associated with ore bodies may yield major clues about the processes of formation of the ores. The directions of remanence of the ores and associated rocks may also be used for dating purpose or for deciphering the tectonic evolution of the region on both a large and small scale. The number of palaeomagnetic observations which have been made on ore bodies is, so far, few, but it is clear that there is a considerable potential in applying palaeomagnetic techniques to a variety of problems associated with the genesis and evolution of ore bodies.

References

Crecelius, E. A., *Carpenter*, R. & *Merrill*, R. T. (1973): Magnetism and Magnetic Reversals in Ferromanganese Nodules. — Earth Planet. Sci. Letters *17*, 391—396.

Hanuš, V. and *Krs*, M. (1963): Palaeomagnetic dating of Hydrothermal Mineralization on Example of Spišsko-gemerské Rudohorie Area. — Czechoslovakia, Rozh. Česk. akad. ved., Sešit *14*, Roč. *73*, pp. 88.

Heirtzler, J. R. *Dickson*, G. O., *Herron*, E. M., *Pitman* III, W. C., & *Le Pichon*, X. (1968): Marine Magnetic Anomalies, Geomagnetic Field Reversals, and Motions of the Ocean Floor and Continents. — J. Geophys. Res., *73*, 2119—2136.

Hood, P. J. (1961): Palaeomagnetic Study of the Sudbury Basin. J. Geophys. Res. *66*, 1235—1241.

Irving, E. (1964): Palaeomagnetism and its application to geological and geophysical problems. — Wiley, N. Y., p. 339.

Ishikawa, Y. & *Syono*, Y. (1963): Order-disorder transformation and reverse thermoremanent magnetism in the $FeTiO_3$-Fe_2O_3 system. — J. Phys. Chem. Solids, *24*, 517—528.

Larochelle, A. (1967): The Palaeomagnetism of the Sudbury Diabase Dyke Swarm. — Canad. J. Earth. Sci., *4*, 323—332.

Mitchell, A. H. G., & *Gamon*, M. S. (1972): Relationship of porphyry copper and circum-Pacific tin deposits to palaeo-Benioff zones. — Trans. Inst. Min. Metall., B *82*, 10—25.

McElhinny, M. W. (1972): Palaeomagnetism and Plate Tectonics. — Cambridge Univ. Press, pp. 358.

Nagata, T., *Uyeda*, S., & *Akimoto*, S. (1952): Self-reversal of thermoremanent magnetism in igneous rocks. — J. Geomagn. Geoelect., *4*, 22—38.

Néel, L. (1955): Théorie du trainage magnétique des ferromagnétiques au grains fins avec application aux terres suites. — Ann. Géophys., 7, 90—102.

Parks, J. M. (1972): Two-state first-order Markov chain model for geomagnetic reversals. — Geol. Soc. Amer., Abs. with Prog., *4 (7)*, 622.

Sawkins, F. J. (1972): Sulfide Ore Deposits in Relation to Plate Tectonics. — J. Geol., *80*, 377—397.

Symons, D. T. A. (1967a): Palaeomagnetic Evidence on the Origin of the Marquette and Steep Rock Hard Hematite and Geothite Deposits. — Canad. J. Earth Sci., *4*, 1—20.

Symons, D. T. A. (1967b): Palaeomagnetic Evidence on the Genesis of the Hard Hematite Ore Deposits of the Vermilion Range, Minnesota. — Canad. J. Earth Sci., *4*, 449—460.

Tarling, D. H. (1971): Principles and Applications of Palaeomagnetism (Chapman & Hall, London), p. 164.

Tarling, D. H. (1973): Metallic Ore Deposits and Continental Drift. — Nature, *243*, 193—196.

Wilson, R. L. (1970): Permanent Aspects of the Earth's Non-dipole Magnetic Field over Upper Tertiary Times. — Geophys. J. R. Astr. Soc., *19*, 417—437.

Wilson, R. L. (1971): Dipole offset — the time average palaeomagnetic field over the past 25 million years. — Geophys. J. R. Astr. Soc., *22*, 491—504.

The Possibility of Geochemical Provinces in the Ocean Basins

Earl *Ingerson**

Abstract

From early chemical analyses of igneous rocks of ocean basins, particularly those near the margins of the basins, it was postulated that there are two different types of oceanic igneous rocks. The "Pacific series" was characterized as of tholeiitic type and the "Atlantic series" as alkalic with a low alkali-lime index.

As analyses accumulated, particularly from midocean ridges, it became apparent that the postulated sharp division was not valid. Moreover, analyses of rocks from sequences in large volcanic piles like the Hawaiian Islands indicated that the alkalic volcanics are probably derived by differentiation of a tholeiitic magma, which appears to be the fundamental one in *all* ocean basins. This impression is strengthened by studying analyses of basal basalts from mid-ocean ridges in the Indian and Atlantic oceans, from the Hawaiian Islands, the East Pacific Rise, etc. The percentages of major elements overlap completely and it is not possible to select any element or combination of elements that would characterize any particular area.

The same thing, by and large, is true of trace elements, although some authors have postulated geochemical provinces in oceanic areas on the basis of the rather scanty data available. There are larger percentage variations than those of major elements, but these may be due to the use of different analytical techniques by the various laboratories, inadequate or improper sampling methods, submarine alteration, mild metamorphism, etc. Differentiation produces much larger variations, so the only really valid comparison is of analyses of flows of primitive undifferentiated tholeiitic magma.

* Prof. Dr. Earl *Ingerson*, Dept. of Geology, University of Texas, Austin, Texas 78 712, USA.

Most of the analyses available are of samples dredged from ocean margins and ridges, sea mounts, etc. What is needed now is to fill in some of the gaps by analyzing cores of igneous rocks brought up from underneath the cover of sediments on the broad ocean floors and eventually from some of the deeps. Analyses for major and trace elements should be done on the same samples and, particularly for trace elements, the determinations should be done in the same laboratory, or if in different laboratories by the same methods and closely co-ordinated so that the results will be as closely comparable as possible.

Zusammenfassung

„Geochemische Provinzen auf den Ozeanböden"

Aus früheren chemischen Analysen magmatischer Gesteine von den Ozeanböden und hier vor allem solcher, die nahe der Ränder der Ozeanbecken genommen wurden, wurde die Forderung nach zwei verschiedenen Typen ozeanischer magmatischer Gesteine erhoben. Die „Pazifische Reihe" war durch den tholeiitischen Magmentyp, die „atlantische Reihe" als alkalisch mit niedrigem Kalk-Alkali-Index gekennzeichnet.

Sobald mehr Analysen vor allem von den mittelozeanischen Rücken vorlagen, wurde klar, daß diese scharfe Zweiteilung keine Gültigkeit hat. Vielmehr ergaben die Analysen ganzer Gesteinsserien in großen Vulkanbauten wie z. B. den Hawaiischen Inseln, daß die alkalibetonten Effusiva wahrscheinlich durch Differenziation aus tholeiitischem Magma entstanden sind, welches das ursprüngliche Magma in allen Ozeanbecken zu sein scheint. Dieser Eindruck wird durch das Studium der Analysen von Basalten der mittelozeanischen Rücken des Indischen und Atlantischen Ozeans, von den Hawaiischen Inseln, dem Ostpazifischen Rücken und anderen bestärkt. Die Prozentanteile der Hauptelemente decken sich weitgehend und man kann kein Element oder keine Elementgruppe erkennen, die ein bestimmtes Gebiet charakterisieren würde.

Dasselbe gilt im großen und ganzen auch für die Spurenelemente, obwohl einige Autoren trotz der geringen Zahl verfügbarer Daten geochemische Provinzen in ozeanischen Bereichen unterschieden haben. Es gibt hier größere Schwankungen der Anteile als bei den Hauptelementen, was aber auch von den verschiedenen angewandten Bestimmungsmethoden in verschiedenen Laboratorien, unzulänglicher oder ungeeigneter Probenahme, untermeerischen Veränderungen, schwacher Metamorphose etc. herrühren mag. Die Differenziation erzeugt viel größere Unterschiede, sodaß der einzig zulässige Vergleich auf Analysen von Lavaströmen ursprünglichen, undifferenzierten tholeiitischen Magmas beruht.

Die meisten der verfügbaren Analysen stammen von Proben, die von den Rändern der Ozeanbecken, ozeanischen Rücken oder von Guyots gedredgt wurden. Was wir heute anstreben ist, die bestehenden Lücken durch Bohrkernanalysen magmatischer Gesteine, die vom Untergrund der Sedimentbedeckung der breiten Ozeanböden und eventuell auch von den Tiefseegräben stammen, zu füllen. Die Analysen der Haupt- und der Spurenelemente sollte an denselben Gesteinsproben vorgenommen, und die Bestimmungen — dies gilt vor allem für die Spurenelemente — im selben Laboratorium durchgeführt werden, oder wenn dies nicht möglich ist, zumindest mit denselben Methoden und in enger Zusammenarbeit erfolgen, um so gut als möglich vergleichbare Ergebnisse zu erhalten.

Introduction

The idea of chemical inhomogeneities in the upper mantle that may extend into or beneath ocean basins as recognizable geochemical provinces is not new. The nomenclature was different, but the concept goes back at least to the publication of Volume II of J. E. *Spurrs Ore Magmas* in 1923. He used the notion that there are areas in the "stable under-earth" enriched in certain metallic elements to explain both "metallographic" (metallogenic) provinces and metallogenic epochs.

Spurr describes the immense distribution of late Tertiary lavas all around the Pacific, but points out that only in certain restricted areas, such as in parts of Arizona and New Mexico, are there significant copper deposits associated with the lavas. He attributes this to heterogeneities in the material underneath the crust, — the "permanent storehouse of the metals", from which the magmas are generated, or at least where they acquire their abnormal load of ore metals.

Spurr pictures this as the lowermost of three zones that „have to do with ore deposits". The intermediate zone is in the lower part of the crust and is "stable neither as to form nor as to position". In a lower sub-zone of the intermediate zone the magma moves but does not differentiate; the upper sub-zone is one of differentiation.

Differentiation produces concentrations of ore metals, in metallographic provinces, which give rise to ore deposits in the top part of the intermediate zone and the lower parts of the uppermost zone, which is a superficial crust of consolidated rocks, and is stable in form but unstable in position.

Spurr postulates that the heterogeneities in the lowermost layer (our upper mantle) "originated at least far prior to the beginning of our geological record". This picture readily explains the existence of a metallogenic province at a given time, but in order to explain repeated metallogenic epochs that supplied the same element(s) to the same area of the crust at widely separated times it is necessary to assume that the heterogeneities in the upper mantle were not only in fixed positions with respect to each other, but also that they had a fixed relation to the overlying crustal rocks.

For example, there have been four periods of copper mineralization in Arizona, ranging in age from Precambrian to late Tertiary and separated by spans of time from about 50 m. y. to some 500 m. y. *Spurr* pictures only localized movement of magma picking up copper from the copper-rich areas and moving upward into the crust in response to imbalance produced by erosion, telluric pressures, etc.

Large scale, deep acting convection currents in the mantle would destroy the inhomogeneities at the top of the mantle, or at least would remove them from their respective positions subjacent to "metallogenic provinces" in the crust. Likewise, large scale movements of continental masses over, or through the upper part of, the mantle would destroy these relationships.

It appears to be necessary to assume, therefore, either 1) that there have been no large scale relative movements between mantle and crust, at least since very early geologic time or 2) that the "permanent storehouses of the metals" are not in the mantle at all, but in a zone low in the crust where they can maintain more or less constant spatial relations to the overlying units even if the whole system is in motion with relation to the underlying mantle.

It is not difficult to imagine a model of this kind, — heterogeneities near the base of the crust that are enriched in ore-forming metals. They could be in a zone of magma generation and/or in the path of a more primitive magma ascending from the mantle, which could pick up metal from the "storehouse" and on subsequent differentiation yield up at least part of it to form ore deposits.

The situation is quite different in ocean basins. If the present widely accepted hypotheses about sea floor spreading, plate tectonics and continental drift are correct, then vigorous subduction zones at continental margins would preclude extension of the heterogeneities that may serve as the basis of geochemical and metallogenic provinces into or beneath the ocean basins.

Is it were established that the geochemical and/or metallogenetic provinces do extend from continental masses into adjacent ocean basins the reality of the supposed subduction zones could be questioned and a model involving permanent chemical heterogeneities in the upper mantle that transcend continental boundaries would again become a theoretical possibility.

On the other hand, if no such extensions can be demonstrated the question of the existence of geochemical provinces in the ocean basins is a problem that is not directly related to that of similar provinces on the continents and must be approached independently.

If the current popular theory of up-welling-translation-subduction has any validity it will certainly not be possible to identify any "permanent" inhomogeneities in the oceanic crust or the underlying upper mantle, which have existed since early geologic times.

Because there are so many uncertainties about the rate, direction, and depth of circulation of possible convection currents in the mantle, it would be very difficult to set up a model that would enable one to predict the types, locations, extent, duration, etc., of geochemical provinces that might exist in ocean basins.

A more realistic approach would be to demonstrate empirically whether such provinces do exist. If their existence can be proved, then a model constructed to explain their properties might also help in the critical study of some of the hypothetical and controversial mechanisms that have been proposed to explain other features of ocean basins. Since large numbers of samples from many places in the ocean basins are becoming available for study, a logical approach would be to study the question of geochemical provinces in oceanic materials first. Then, if such provinces can be identified and outlined, the problem of their possible correlation with similar provinces on the continental masses can be pursued subsequently.

The suggested empirical study of the geochemistry of ocean basins might consist of three stages for the present and near future:

1. Compile, study, and plot all usable analyses, both for major and trace elements, that have been made of igneous rocks from all of the principal ocean basins.

2. Analyze all available appropriate samples from the current deep drilling programs. When preliminary results are available perhaps the scientific staffs could be prevailed upon to modify plans as to depth of drilling and/or details of location of holes in the interest of contributing more significantly to the geochemical studies.

3. The currently planned drilling "legs" will provide only a very loose network of sample locations. Many of the drill holes do not penetrate to the igneous rocks, or they provide only a little weathered material. If steps 1 and 2 appear to warrant it, funds might be obtained to mount a special drilling program to fill in the critical areas, re-enter holes to get deeper and better samples, etc.

In considering step 1, compilation and interpretation of existing data, the problem of defining "usable analyses" immediately arises. There is no point to plotting all available analyses of igneous rocks, or even all extrusive rocks, because differentiation has proceeded to different extents and possibly in different directions, in different parts of the ocean basins. Even in a relatively compact "province" like the Hawaiian Islands *(Macdonald* 1949), the volcanic centers, even on a single island, show marked differences in degree of differentiation, with attendant large differences in amounts and ratios of both major and trace elements.

The problem is further complicated by the probability that the primary tholeiitic magma can differentiate either in the direction andesite-trachyte-rhyolite, or to produce alkalic end products such as nepheline basalt *(Bowen,* 1928; *Engel* & *Engel,* 1964a). Therefore, in comparing analyses of rocks for the pupose at hand we should, ideally, select samples from different areas that represent differentiation in the same sequences (calcalkalic or alkalic) and to the same degree. Moreover, other factors such as submarine alteration and mild metamorphism may change the composition, especially with respect to trace elements.

Since it is not possible to match degrees of differentiation accurately, or to evaluate trace element loss, or gain, during submarine alteration the logical samples to use would be fresh, unmetamorphosed ones of primary ("primitive", undifferentiated) tholeiitic basalt. It is easy to recognize, either in the samples themselves or in most published descriptions, the cumulates like oceanite or picrite-basalt and differentiates to which names other than "basalt" are given. There is still the problem of which "basalt" analyses to use for comparison; how much olivine does a "primitive" tholeiitic basalt contain? *Tröger* (1935) has given 7 % for the type tholeiite; and the A. G. I. Glossary says a tholeiitic magma is one that "contains little or no olivine".

Perhaps the best we can do on the initial attempt is to use analyses of rocks called "tholeiitic basalts" or simple basalts, unless there is evidence in the descriptions that they contain excessive cumulate olivine or other irregularities that make them suspect. Presumably the authors and analysts will have selected the most representative and freshest material available.

There is another possible difficulty and there appears to be no way to evaluate it at the moment. The magma appears to be generated by differential melting of mantle material and the *degree of melting* at different places and at different times may produce magmas with somewhat different compositions, even without inhomogeneities in the mantle, or any differentiation subsequent to generation.

Most of the analyses made thus far are from lavas of volcanic islands and samples that have been dredged from seamounts and other relatively thick sequences. These are the places where differentiation has been most active, so the analyses must be selected with care. On the great oceanic plains the flows are undoubtedly fewer and thinner and it is probable that they are more nearly representative of the primitive tholeiitic magma. For that reason the samples obtained by drilling these areas should be interesting and important.

For the purpose at hand no attempt has been made to glean from the literature *all* of the acceptable analyses of oceanic tholeiitic basalts, but to select representative analyses and averages for widely separated areas in the three principal oceans where enough samples have been collceted to give comparisons of the over-all compositions of the primary (primitive) tholeiitic rocks in the different areas.

The following were selected:
1. Mid-Atlantic Ridge
2. East Pacific Rise
3. Hawaiian Islands
4. Indian Ocean Ridges

Rocks from island arcs were not considered because they resemble continental rocks more than oceanic ones in type, association, etc. Two hundred new analyses of Canary Island "basaltic rocks" *(Ibarrola,* 1969) were neglected, because they are the result, in part, of alkalic differentiation, which produced some feldspathoid-bearing varieties; many others are high-olivine basalts, alkalic olivine basalts, etc.

The localities chosen span some 240⁰ of longitude and at least 70⁰ of latitude. As to petrographic type the rocks are described as basalt, tholeiite, tholeiitic basalt, olivine-bearing basalt, etc. An attempt was made to choose analyses of rocks with less olivine than the 7 % mentioned for tholeiite by *Tröger* and most of them have less than 5 % olivine. The assumption is that more olivine indicates a cumulate rather than a primary igneous rock. No analyses were selected where the presence of a feldspathoid was indicated or where the alkali-lime index suggested alkalic affinities.* However, in many cases it was not possible to be sure of mineral composition, because descriptions gave only norms, or no mineralogical composition at all, — only rock names.

Analyses showing more than 2.5—3 % water were rejected. Larger amounts change the dry weight percentages of the other constituents significantly. Moreover, high water content indicates weathering or some other type of alteration, which may have an effect on concentration of some elements.

Major Elements

Table 1 gives the ranges of major element composition for tholeiitic basalts from the four selected areas mentioned above. A glance at the table shows that there are very considerable overlaps for most of the components and that many of them overlap almost completely. The Hawaiian basalts appear to be consistently high in TiO_2 and low in Al_2O_3. This is due in part to the fact that the figures for Hawaii are from averages for the various centers,whereas the other columns represent ranges for all the individual analyses. If we take the individual analyses for Hawaii the overlaps for TiO_2 and Al_2O_3, as well as for some of the other components, are larger.

Table 2 gives some averages that have been suggested for the tholeiitic basalts in the areas under consideration and two more general averages. It should be emphasized that these averages are not comparable in numbers or selection of samples, areas covered, volumes represented, etc.; they are just the averages of the analyses available. Nevertheless, the averages are remarkably consistent for values obtained in the way they were. The larger deviation for the Hawaiian samples (eight of eleven components outside the range of the other averages) is due to the fact that a great many of the samples were olivine-bearing tholeiites, which explains the low values for silica, alumina, and Na_2O, and the high ones for FeO, MgO and TiO_2.

* Except trace element contents of some rocks from Polynesia (Table 3), which are included for comparison. These results on alkalic rocks are not included in any of the averages and are not used to determine total spread for the rock types under consideration.

Table 1
Major Elements in Tholeiitic Basalts

	Mid-Atl. Ridge[1]	Ind. Ocean Ridges[2]	E. Pac. Rise[3]	Hawaii[4]
SiO_2	47.11—50.91	49.52—51.39	49.13—50.42	48.39—50.66
TiO_2	0.72—2.03	0.97—1.37	0.99—2.18	2.09—2.86
Al_2O_3	14.03—20.07	15.88—21.09	14.85—17.42	13.25—14.95
Fe_2O_3	0.85—2.76	0.94—4.14	1.35—4.30	1.88—4.83
FeO	4.35—9.89	7.09—7.88	4.54—10.24	6.83—9.75
MgO	6.00—10.29	5.47—8.65	5.82—8.37	7.70—9.17
CaO	10.44—14.10	10.86—13.22	10.45—12.83	9.48—10.60
Na_2O	2.10—3.15	2.64—3.51	2.37—3.24	1.84—2.50
K_2O	0.08—0.77	0.14—0.25	0.11—0.38	0.13—0.54
P_2O_5	0.10—0.19	0.04—0.14	0.04—0.28	0.19—0.30
MnO	0.09—0.21	0.09—0.19	0.11—0.22	0.12—0.18

[1] *Aumento* (1968, 1969); *Engel & Engel* (1964a); *Muir & Tilley* (1964); *Nicholls, Nalwalk & Hays* (1964).
[2] *Engel & Fisher* (1969); *Engel, Fisher & Engel* (1965).
[3] *Engel & Engel* (1963, 1964b).
[4] *Macdonald & Katsura* (1964).

Table 2
Major Element Averages — Tholeiitic Basalt

	Mid-Atl. Ridge[1]	Ind. Ocean Ridge[2]	Carls-berg Ridge Ind. O.[3]	East Pacific Rise[4]	Hawaii[5]	Ocean Tholeiites[6]	Ocean Floor Basalts[7]
SiO_2	49.78	50.65	51.81	50.25	49.36	49.94	49.61
TiO_2	1.29	1.36	1.88	1.56	2.50	1.51	1.43
Al_2O_3	16.92	17.09	15.56	16.09	13.94	17.25	16.01
Fe_2O_3	1.94	1.66	3.56	2.72	3.03	2.01	
FeO	7.32	7.00	6.39	7.20	8.53	6.90	11.49
MgO	8.18	7.49	7.10	7.02	8.44	7.28	7.84
CaO	11.34	11.52	9.35	11.81	10.30	11.86	11.32
Na_2O	2.77	2.82	3.87	2.81	2.13	2.76	2.76
K_2O	0.16	0.18	0.11	0.20	0.38	0.16	0.22
P_2O_5	0.14	0.08	0.20	0.15	0.26	0.16	0.14
MnO	0.16	0.15	0.17	0.19	0.16	0.17	0.18

[1] *Engel, Fisher & Engel* (1965).
[2] Ibid.
[3] Ibid.
[4] Ibid.
[5] *Macdonald & Katsura* (1964).
[6] *Engel, Engel & Havens* (1965).
[7] *Cann* (1971).

The striking similarity in tholeiitic basalts over very large areas in ocean basins has led to the belief that they represent in composition the predominant if not the only magma generated beneath the ocean floors (e. g. *Engel & Engel,* 1964a). Even those who believe that alkalic magmas are generated independently agree that the tholeiitic one is very much more abundant and probably more "fundamental".

Bowen (1928) believed that both calc-alkalic and alkalic rocks could be formed by differentiation of this fundamental magma and developed a model, largely on experimental evidence, of how it could happen. A model like this is now widely accepted. There are compelling lines of evidence other than Bowen's experimental demonstration. The alkalic rocks in big oceanic piles like the Hawaiian Islands commonly occur near the tops of the piles; that is, they are very late in the sequence and they ascend through the same channels that gave rise to the earlier sub-alkalic rocks.

Also, trace elements that are not readily fractionated early in the sequence, such as Co, Ga, V, Y, Yb, etc., show almost identical concentrations in tholeiitic and alkalic basalts. It is highly improbable that this would be the case if the two kinds of rocks were produced from different kinds of magma that originated entirely independently.

Cann (1971) made a detailed study of major element variations in ocean-floor basalts and concluded that these can be explained by "precipitation of calcic plagioclase, forsteritic olivine and augite" and probably to a lesser extent by "crystal-liquid processes involving amphibole".

It would be possible to cite many more opinions and to give much more data supporting the essential uniformity of primary tholeiitic basalts of the great ocean basins. Two or three more examples should suffice for the present. *Kay, Hubbard & Gast* (1970) conclude that observed variations in the chemical composition of oceanic ridge volcanic rocks can be ascribed to shallow differentiation dominated by plagioclase and olivine.

Bonatti & Fisher (1971) have studied volcanic rocks on and at a distance from oceanic ridges and they conclude that "the conditions of magma generation are similar beneath ridges and away from them".

Nicholls & Islam (1971) appear to accept the essential uniformity of the major element composition of tholeiitic basalts from the ocean floor, but state that variations in trace elements are much more marked than those in major element contents. Theirs is the only positive statement encountered that "Geochemical provinces exist in oceanic areas just as they do in continental regions." See below, under *Trace Elements,* the discussion of these ideas.

Trace Elements

Trace element ranges for tholeiitic basalts from the same areas as those selected for major elements are shown in Table 3. The last two columns show values for some Polynesian basalts with alkalic tendencies and the effects of different degrees of silica saturation, respectively. As would be expected, the undersaturated alkalic rocks (normative Ne) are high in such things as Sr, Ba, Rb and Zr, whereas the undersaturated calcalkalic rocks (normative Ol) are high in Cr and Ni. Co, Ga, Sc, V and Y do not show large or systematic variations, even in alkalic rocks or those with variable silica saturation.

Table 3
Trace Elements in Tholeiitic Basalts

T. E.	Mid-Atl. Ridge[1]	Indian Ocean Ridges[2]	East Pacific Rise[3]	Hawaii[4]	Poly-nesia[5]*	Dif. Silica Satur.[6]
Ba	20—200	6—96	6—25	50—125	200—550	14—498
Co	22—100	26—110	28—38	30—60	25—60	25—42
Cr	40—700 (3000)	65—420	160—460	300—500	125—300	67—700
Cu	45—100	62—110	64—87	100—200	—	30—100
Ga	10—25	12—38	14—20	15—30	20—25	7—22
Li	2—10	—	3—6 (21)	1—10	10—20	2—11
Ni	70—220 (1000)	85—170	58—140	70—350	120—320	51—220
Rb	5—28	4—8	10	—	5—40 (275)	1—33
Sc	10—65	20—80	30—56	10—30	—	15—61
Sr	20—320	85—240	97—120	350—600 (1000)	500—1000 (1800)	130—815
V	70—420	120—440	170—440	200—400	150—280	220—350
Y	10—60	33—39	21—60	20—30	20—25	10—54
Yb	3—6	3—4	2—7	—	—	—
Zn	51—100	—	—	—	—	35
Zr	30—160	61—165	44—150	50—200	175—230	45—333

* Alkalic tendency.
[1] *Aumento* (1968, 1969); *Engel, Engel & Havens* (1965).
[2] *Muir & Tilley* (1964); *Engel & Fisher* (1969); *Nicholls & Islam* (1971).
[3] *Engel, Engel & Havens* (1965).
[4] *Nockolds & Allen* (1956); *Wager & Mitchell* (1953).
[5] *Nockolds & Allen* (1954).
[6] *Nicholls & Islam* (1971).

If we do not consider the analyses of the Polynesian rocks of alkalic affinity then all of the trace element results from the other four areas show just as extensive overlaps as do the major element compositions of rocks from these same areas. Very many fewer trace element analyses are represented in Table 3 than the major element analyses of Table 1. If we also consider that slight contamination or micro-inhomogeneities, differences in methods, techniques and interpretation and other vagaries of analyses of rocks for trace elements can easily result in differences of an order of magnitude in reported results, there appears to be no more reason to postulate geochemical provinces in ocean basins on the basis of trace elements than from the major element compositions.

Most of the authors of papers reporting trace element results do not concern themselves with problems of geographic differences. They are more concerned with patterns of trace element distribution in a given area, and the bearing it has on questions of magma origin, and mechanism of development of the series (singular or plural) of rocks under consideration. For example, concerning the rocks of the Mid-Atlantic Ridge *Muir* & *Tilley* (1964) merely say, "It is clear that a much more intensive collection and study of the rocks ... is necessary". *Wager* & *Mitchell* (1953) from a study of trace elements in Hawaiian rocks conclude that "The results are in conformity with the view that the Hawaiian series are essentially the result of fractional crystallization".

Gast (1971) studied trace elements in ocean ridge basalts from three widely separated localities. He mentions no geographic differences but concludes that variations in trace element contents are explained by crystallization of plagioclase and olivine from a magma whose over-all chemistry is determined by partial melting of mantle material at depths of 15—25 km.

Nicholls & *Islam* (1971) discuss a number of factors that may cause apparent geographic variations in trace element contents of oceanic basaltic rocks. These probably explain most of the observed variations but Nicholls and Islams believe that after all these have been considered "... there remain variations in trace element contents of otherwise comparable basalts from different parts of the ocean floor, which appear to represent real variations in the trace element contents of the erupted basaltic magmas".

When we look at the discussion of individual trace elements, however, there does not seem to be a strong case for significant geographic differences. For practically all of the elements discussed speciafically they say, "there are too few data", or there is "no real evidence of regional variations" or "the results are erratic", etc. Barium, to the discussion of which more than a page is devoted, appears to be the element on which Nicholls and Islam depend most heavily in postulating geochemical provinces.

On general grounds it appears that positive results on more than one element are needed to support such a fundamental postulate. And barium may be a particularly unfortunate element for such a role. Its concentration changes very rapidly and by at least an order of magnitude during differentiation in the basaltic range *(Wager* & *Mitchell*, 1951, Fig. 6). This is undoubtedly the explanation of some of the high Ba values of individual analyses.

Unless one could be *sure*, therefore, that basalts being compared from different localities are at almost *exactly* the same stage of differentiation, the barium concentrations would not be significant in delineating geographic differences.

Conclusion

From early chemical analyses of igneous rocks of ocean basins, particularly those near the margins of the basins, it was postulated that there are two different types of oceanic igneous rocks. The "Pacific series" was characterized as of tholeiitic type and the "Atlantic series" as alkalic with a low alkali-lime index.

As analyses accumulated, particularly from mid-ocean ridges, it became apparent that the postulated sharp division was not valid. Moreover, analyses of rocks from sequences in big volcanic piles like the Hawaiian Islands indicated that the alkalic volcanics are probably derived by differentiation of a tholeiitic magma, which appears to be the fundamental one in *all* ocean basins. This impression is strengthened by studying analyses of basal basalts from mid-ocean ridges in the Indian and Atlantic oceans, from the Hawaiian Islands, the East Pacific Rise, etc. The percentages of major elements overlap completely and it is not possible to select any element or combination of elements that would characterize any particular area.

The same thing, by and large, is true of trace elements. There are larger percentage variations, but these may be due at least in part to the use of different analytical techniques by the various laboratories.

From thoughtful consideration of these data, some of which are tabulated above, it does not appear that current available information can justify any strong stand in favor of the idea of geochemical provinces in ocean basins. I hope that I have not appeared prejudiced in favor of this point of view. I began this study fully expecting to find evidence of such provinces and thought that the principal problem would be, not the question of their existence, but rather their delineation and characterization.

Most of the analyses available now are of samples dredged from ocean margins and ridges, seamounts, etc. What is needed now is to fill in some of the gaps by analyzing cores of igneous rocks brought up from underneath the cover of sediments on the broad ocean floors and eventually from some of the deeps. This would be essentially stages two and three of *Suggested Studies*, above.

Analyses for major and trace elements should be done on the same samples and, particularly for trace elements, the determinations should be done in the same laboratory or if in different laboratories by the same methods and closely co-ordinated so that comparisons of analyses would be as meaningful as possible. It might be helpful to reanalyze (again, especially for trace elements) splits of some samples previously analyzed and reported. This could indicate the magnitude of some variations in results which are not due to geographic differences.

References

Aumento, F. (1968): The Mid-Atlantic Ridge near 45⁰ N. II. Basalts from the area of Confederation Peak. — Canadian Jour. of Earth Sci., V. *5*, No. 1, pp. 1—21.
— (1969): Diorites from the Mid-Atlantic Ridge at 45⁰ N. — Science, *165*, No. 3898, pp. 1112—1113.

Bonatti, E. & D. E. *Fisher* (1971): Oceanic basalts; chemistry vs. distance from oceanic ridges. — Earth and Plan. Sci. Letters, *11,* p. 307.

Bowen, N. L. (1928): The evolution of the igneous rocks. — Princeton University Press, 332 pages.

Cann, J. F. (1971): Major element variations in ocean-floor basalts. — Phil. Trans. Roy. Soc. London A., *268,* No. 1192, pp. 495—505.

Engel, A. E. J. & C. G. *Engel* (1964a): Composition of basalts from the Mid-Atlantic Ridge. — Science, *144,* No. 3624, pp. 1330—1333.

— (1964b): Igneous rocks of the East Pacific Rise. — Science, *146,* No. 3643, pp. 477—485.

Engel, A. E. J., C. G. *Engel* & R. G. *Havens* (1965): Chemical characteristics of oceanic basalts and the upper mantle. — Bull. Geol. Soc. Amer., 76, No. 7, pp. 719—734.

Engel, C. G. & A. E. J. *Engel* (1963): Basalts dredged from the Northeastern Pacific Ocean. — Science, *140,* No. 3573, pp. 1321—1324.

Engel, C. G. & R. L. *Fisher* (1969): Lherzolite, anorthosite, gabbro and basalt dredged from the Mid-Indian Ocean Ridge. — Science, *166,* No. 3909, pp. 1136—1141.

Engel, C. G., R. L. *Fisher* & A. E. J. *Engel* (1965): Igneous rocks of the Indian Ocean floor. — Science, *150,* No. 3696, pp. 605—609.

Gast, P. W. (1971): Dispersed element chemistry of oceanic ridge basalts (abs.). — Phil. Trans. Roy. Soc. London. A., *268,* p. 467.

Ibarrola, E. (1969): Variation trends in basaltic rocks of the Canary Islands. — Bull. Volc., *33,* pp. 729—777.

Kay, R., N. J. *Hubbard* & P. W. *Gast* (1970): Chemical characteristics and origin of oceanic ridge volcanic rocks. — Jour. Geoph. Res., *75,* pp. 1585—1613.

Macdonald, G. A. (1949): Hawaiian petrographic province. — Bull. Geol. Soc. Amer., *60,* pp. 1541—1595.

Macdonald, G. A. & T. *Katsura* (1964): Chemical composition of Hawaiian lavas. — Jour. Petr., *5,* Part 1, pp. 82—133.

Muir, I. D. & C. E. *Tilley* (1964): Basalts from the northern part of the rift zone of the Mid-Atlantic Ridge. — Jour. Pet., *5,* Part 3, pp. 409—434.

Nicholls, G. D. & M. R. *Islam* (1971): Geochemical investigations of basalts and associated rocks from the ocean floor and their implications. — Phil. Trans. Roy. Soc. Lond. A., *268,* No. 1192, pp. 469—491.

Nicholls, G. D., A. J. *Nalwalk* & E. E. *Hays* (1964): The nature and composition of rock samples dredged from the Mid-Atlantic Ridge between 22⁰ N and 25⁰ N. — Marine Geol., *1,* pp. 333—343.

Nockolds, S. R. & R. *Allen* (1954): The geochemistry of some igneous rock series: Part II. — Geochim. Cosmochim. Acta, *5,* No. 6, pp. 245—285.

— (1956): The geochemistry of some igneous rock series: Part III. — Geochim. Cosmochim. Acta, *9,* Nos. 1/2, pp. 34—77.

Tröger, W. E. (1935): Spezielle Petrographie der Eruptivgesteine. — Verlag der Deutschen Mineralogischen Gesellschaft e. V., Berlin, 360 pages.

Wager, L. R. & R. L. *Mitchell* (1951): The distribution of trace elements during strong fractionation of basic magma. — Geochim. Cosmochim. Acta, *1,* pp. 129—208.

— (1953): Trace elements in a suite of Hawaiian lavas. — Geochim. Cosmochim. Acta, *3,* No. 5, pp. 217—223.

Über den Unterschied der erzbildenden Vorgänge in den Kontinenten und den Ozeanen

Yu. M. *Scheinmann**

Der Unterschied zwischen ozeanischer und kontinentaler Metallogenese besteht vorwiegend in der Tatsache, daß im ozeanischen Bereich die an saure Magmen geknüpften Lagerstätten fehlen. Man erklärt dies zumeist mit der Abwesenheit saurer Magmen in der ozeanischen Kruste.

Die Verhältnisse sind aber komplizierter. Wir wissen heute, daß man in den mittelozeanischen Schwellen geschichtete Intrusionen gefunden hat, in denen nicht nur Gabbro, sondern auch Granodiorit und Natrium-reicher Granit vorkommen. Aber diese sauren Gesteine bzw. ihre Muttermagmen sind Erz-steril. Die Intrusionen in der mittelatlantischen Schwelle im Bereich des 45⁰ N können als gutes Beispiel dafür dienen (*Aumento* et al.).

Der Hauptunterschied zwischen ozeanischen und kontinentalen Gesteinen ist nicht so sehr durch den Kali-Gehalt allein, sondern durch das Verhältnis Na_2O zu K_2O bestimmt. Dabei sind zumeist die primären Gesteine zu betrachten, da sie am besten die ursprüngliche Zusammensetzung der Magmen widerspiegeln. Bei einer solchen Betrachtung stellt sich heraus, daß die basaltischen, aber auch die andesitischen und noch saureren Magmen ein Na_2O/K_2O-Verhältnis zwischen 2 und 4 haben, die ozeanischen Magmen dagegen ein solches zwischen 7 und 30. Die geographische Grenze dieser Gesteinsprovinzen liegt immer an der Grenze der ozeanischen Kruste, im Falle der Inselbögen an deren ozeanischer Seite.

So kommen wir zu dem Schluß, daß nicht das Fehlen der sauren Magmen im Ozean der Grund für den Unterschied ist, sondern der Mangel an Gesteinen der Kali-Reihe. Bekanntlich ist die Kali-Reihe an die Kontinente gebunden. Sie fehlt bei den ozeanischen Gesteinen völlig. Kürzlich haben verschiedene Autoren (*Ronov* & *Yaroshevsky*, *Scheinmann*, *Scheinmann* & *Bazhenova*) gezeigt, daß die Menge an Kali in der kontinentalen Kruste so hoch ist, daß es nötig sein würde, dieses Kali aus einer Tiefe von 1000—1500 km aus dem Mantel in die Kruste zu übertragen, Natrium aber nur aus 100 bis 200 km Tiefe.

* Prof. Dr. Yury M. *Scheinmann*, Inst. für Geophysik der Erde, Gruzinskaya 10, Moskwa D-242, USSR.

Das bedeutet, daß der Kali-Gehalt der kontinentalen Kruste nicht durch einfache Aufschmelzung des Mantels erklärt werden kann. Wollte man dies annehmen, so müßte der Mantel im Bereich einer mindestens 1000 km dicken Schale partiell aufgeschmolzen worden sein. Aber eine solche geringfügige Schmelzfraktion von nur 4 % wird keine Möglichkeiten finden, aus der kristallinen Substanz des festen Mantels nach oben zu entweichen und überdies kennen wir keinerlei Schmelzprozesse in der Tiefe von mehreren 100 km.

Also müssen wir eine andere Ursache für den hohen Kali-Gehalt in der kontinentalen Erdkruste suchen. Ich glaube, daß die beste Lösung des Problems in der Anwendung der Hypothese *Artyushkov's* liegt, deren Inhalt darin besteht, daß die Differentiation der Erdmaterie noch nicht zu Ende gekommen ist. Noch heute differenziert sich an der Grenze Erdkern und Erdmantel der Urstoff in die dichte Kernmasse und einen leichteren Silikatrest. Der letztere muß wegen seiner geringeren Dichte aufsteigen.

Solche aufsteigende Ströme bringen immer wieder neue Massen und neue Wärmemengungen in den Bereich der Astenosphäre. Dadurch werden immer neue Mengen von Kali und seinen Begleitern der Kruste zugeführt *(Scheinmann* 1972). Im Bereich der Ozeane fehlen derartige Prozesse, dadurch ist die Erdkruste im etwa ursprünglichen Zustand geblieben.

Folglich muß man annehmen, daß unter den Ozeanen auch keine aufsteigenden Ströme und somit auch kein Kaliaufstieg vorhanden ist.

So ist es eine logische Konsequenz, wenn wir annehmen, daß zusammen mit dem Kali eine Reihe von Elementen aus der Tiefe kommt.

Damit kommen wir zur Hypothese, daß die Mannigfaltigkeit der Erze in den Kontinenten durch die tiefen, vom Unter-Mantel herkommenden Differentiationsströme bestimmt wird und daß diese Erzmannigfaltigkeit im Stoff der ozeanischen Kruste „keinen Zutritt" gefunden hat.

Abstract

"About the difference of ore-forming processes in continental and oceanic provinces"

Not only the basaltic rocks but also their acid differenciates in the sphere of the oceanic crust are barren with regard to the forming of ore deposits. The essential difference between oceanic and crustal magmas is the much higher Na_2O/K_2O ratio of the former. If one considers the great quantity of potassium in continental rocks derived from the mantle, this would imply a melting of the mantle to a depth of about 1500 km. There is no evidence for that. An explanation might give the hypothesis of E. V. *Artyushkov* according to which the differentiation continues at the border core/mantle of the earth until present time.

During this process the light K-rich differentiation products to which the ore-forming elements are bound, ascend. These ascending substances are to be found beneath the continental crust only.

References

Artyushkov, E. V. (1970): Density Differentiation on the Core-Mantle Interference and Gravity Convection. — Physics of the Earth and Planetary Interior *2*, n 5.

Aumento, F. (1969): Diorites from the Mid-Atlantic Ridge at 45⁰ N. — Science, *165*, n 3898, Sept.

Aumento, F. & *Lancarevic*, B. D. (1969): The Middle-Atlantic Ridge near 45⁰ N. — Bauld Mountain. Canad. J. Earth Sci. *6*, n I.

Ronov, A. B. & *Yarochevsky*, A. A. (1969): Chemical composition of the Earth Crust. — "Earth Crust and Upper Mantle", Am. Geoph. Union Monogr. n. *13*.

Scheinmann, Yu. M. (1972): Različie materikovoi i okeaničeskoi litosferi i differenziazia zemli. — Geotektonika *6*.

Scheinmann, Yu. M. & *Baženova*, T. H. (1972): Obštegeologičeskoe značenje nekotorich čert sostava basaltov okeana i materika. — Bull. Mosk. Obšt. Isp. prirodi otdel. geol. *3*.

Metallogenesis and Distribution of Elements Around the Zones of Subduction

E. *Szádeczky-Kardoss*[*]

Abstract

According to a new model of plate tectonics most magmatic rocks derive by the partial melting of the sedimentary cover of the subducted plates. The distribution of the minor elements and the magmatic ore formation are controlled by the correlated processes. Geochemical correlation betweeen ore minerals and conjugated rocks indicates that the heavy metallic cations of the hydrothermal veins are supplied mostly by magmatic solutions, while their sulfidic and related quasianions (S, Se, Te, As, Sb, with some volatile cations e. g. Hg) derive mainly from the (meta) clays of continental origin. (Average S content in SO_3: volcanics 0,03 %, geosynclinale clays and carbonate rocks 0,11 and 0,03 %, continental clays and carbonate rocks 0,43 and 2,36 %.) The investigation of the flow system in the ore veins indicates also that hydrothermal sulfide ore minerals are mostly precipitates of the anion forming elements of descendent-hemiascendent solutions by the heavy metallic ions of ascendent solutions. The pre- and postmetallogenic propylitization and other (clay, silica etc.) wall rock alteration products related to these ore veins are characterized, too, by the volatile elements of sedimentary origin. According to the lower content of sulfur in the oceanic and geosynclinale clays, the hydrothermal ores of the orogenic belts (Lahn-Dill type etc.) form mostly oxydic mineralizations. Both cations and anions of the pegmatitic-pneumatolytic ore formations are of magmatic origin. According to the low sulfur content of the magma they form mostly oxide or silicate and carbonate minerals. They crystallize by the cooling of the magmatic solutions, according to the much higher geothermal gradients during their formation than those of the hydrothermal ores.

The mobilized series of the elements differed considerably for the melts and aquaeous solutions — as computed from the ratio of their average contents

[*] Prof. Dr. Elemér *Szádeczky-Kardoss,* Hungarian Academy of Sciences, Muzeum Körut 4/A, Budapest VIII, Hungary.

in the granites and the crust, e. g. from that in the clays and the crust — corroborate the above statements, too.

Many transitory types exist between these two extreme types of postmagmatic ore formation, e. g. the tin-boron-mineralization accompanied by sulfidic ores.

Examples of subduction zones were given from the Pannonian area, with spezial reference to the Carpathian Arch.

Zusammenfassung

„Metallogenese und Verteilung von Elementen in Subduktionszonen"

In Übereinstimmung mit neuen Auffassungen über die Plattentektonik entstehen die meisten magmatischen Gesteine durch partielles Schmelzen der Sedimentbedeckung unterschobener Platten. Die Verteilung der Nebenelemente und die magmatische Erzbildung werden durch voneinander abhängige Prozesse gesteuert. Die geochemischen Wechselbeziehungen zwischen Erzmineralen und zugehörigen Gesteinen zeigen, daß die Kationen der Schwermetalle der hydrothermalen Gänge meist aus magmatischen Lösungen stammen, während deren sulfidische und verwandte Quasi-Anionen (S, Se, Te, As, Sb, auch einige flüchtige Kationen wie z. B. Hg) hauptsächlich kontinentaler Herkunft (Tone und Metatone) sind. (Der durchschnittliche S-Gehalt als SO_3 beträgt in Vulkaniten 0,03 %, in geosynklinalen Tonen und Karbonatgesteinen 0,11 und 0,03 % und in kontinentalen Tonen und Karbonatgesteinen 0,43 und 2,35 %.) Die Untersuchung der Fließverhältnisse in den Erzgängen zeigt überdies, daß hydrothermale sulfidische Erzminerale meist durch Schwermetall-Ionen aszendenter Lösungen aus dezendenten bis hemi-aszendenten Lösungen, die die Elemente der Anionen enthalten, ausgefällt werden. Propylitisierung vor und nach der Vererzung sowie andere Nebengesteinsveränderungen (Vertonung, Silizifizierung, etc.) in Zusammenhang mit der Vererzung werden durch flüchtige Elemente sedimentärer Herkunft charakterisiert. Entsprechend des niedrigen Schwefel-Gehaltes der ozeanischen und geosynklinalen Tone liegen die hydrothermalen Erze der Orogengürtel meist in oxydischer Form vor (Lahn-Dill-Typus u. a.). In den pegmatitisch-pneumatolytischen Erzformationen sind sowohl Kationen als auch Anionen von magmatischer Herkunft. Entsprechend dem niedrigen Schwefel-Gehalt des Magmas werden meist Oxyde oder Silikate und Karbonate gebildet. Infolge des wesentlich höheren geothermalen Gradienten gegenüber den hydrothermalen Erzen kristallisieren sie als Folge der Abkühlung der magmatischen Lösungen aus.

Die Mobilisationsreihen der Elemente unterscheiden sich für Schmelzen und wässrige Lösungen beträchtlich, wodurch die oben gemachten Feststellungen weiter untermauert werden.

Zwischen diesen beiden extremen Arten der postmagmatischen Erzbildung gibt es eine Reihe von Übergängen, wie z. B. die Zinn-Bor-Vererzung, die von sulfidischen Erzen begleitet wird.

Beispiele von Subduktionszonen werden für das Pannonische Gebiet mit besonderem Hinweis auf den Karpatenbogen gegeben.

1. Two main types of subduction. Upper Cretaceous-Tertiary plate tectonics of the Carpathian-Pannonian area.

In the Circumpacific belt mostly broad oceanic plates are drifting either parallel or slightly divergently. Hence they are not hindered in their movements by each other.

The Mediterranean microcontinents generated by the (partly horizontal) confrontation of two continents are drifted often convergently together with the narrow oceanic stripes of relatively short lifetime between them. Thus, the microcontinental plates are mutually hindered in their movements and become dissected into small sections moving with different velocity. Minimum velocities occur at the edges of the main plate sections resulting in a garlande structure (fig. 1). The accreting margins of the small oceanic plates gradually merge by the subduction into the consuming type and the abyssal sediments mostly vanish.

Circumpacific type

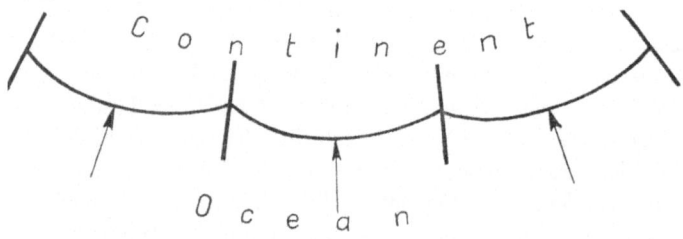

Mediterranean type

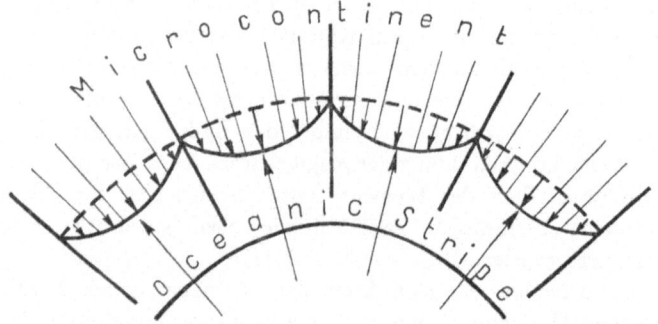

Fig. 1. The Circumpacific and the Mediterranean type of subduction.

Particularities of the mediterranean microcontinents was studied in the Carpathian-Pannonian area. First the sutures of the young Cretaceous and Tertiary subductions *(Author, 1973a)* were traced using more than twenty different methods *(Author 1973b)* and the following system of subductions was found for the Pannonian-Carpathian area (fig. 2 and 3).

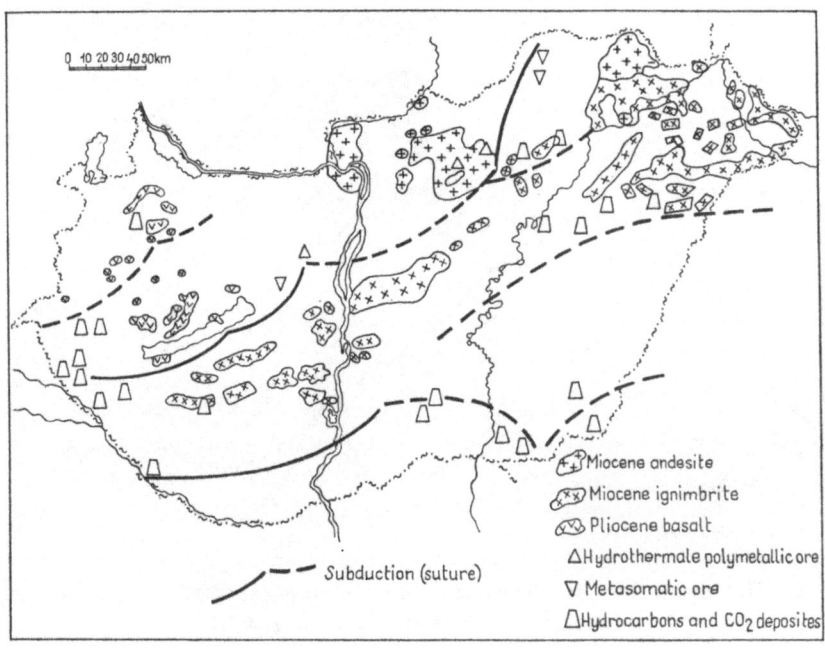

Fig. 2. The suture lines of subductions and their relation with igneous rocks, ores and hydrocarbons in the Pannonian Basin.

1. The Carpathian Klippenzone suture forms on the surface mainly a rock-mélange. It dips convergently towards the Pannonian Basin. The dip angle is mostly flat (about 50^0), in the East it is steeper. The continuation of this subduction in the Alps is presumable (also according to *Dewey* & *Bird*).

2. The marginal subduction of the Hungarian Central Mountains appears on the surface as a mélange ("Darno line"); it is, however, partly covered by the Neogene of the Pannonian Basin. It dips mostly steeply ($70—80^0$) towards the NW. Southwestwards it continues in the Judicaria-Insubria line. Northeastwards it is presumably perpendicularly deflected into the Szamos-Somes-line, described by *Stille*.

3. The subduction at the southern border of Mecsek Mts dipping $\pm$ northwards proceeds eastwards under the Great Hungarian Plain and reaches the surface again in the Maros-Mures-belt of Transylvania forming there a Klippen-belt with mélange.

The presumed continuation of this system in the Alps, Dinarids and Balkans is delineated schematically in fig. 3, mostly according to the volcanic derivatives of the subduction.

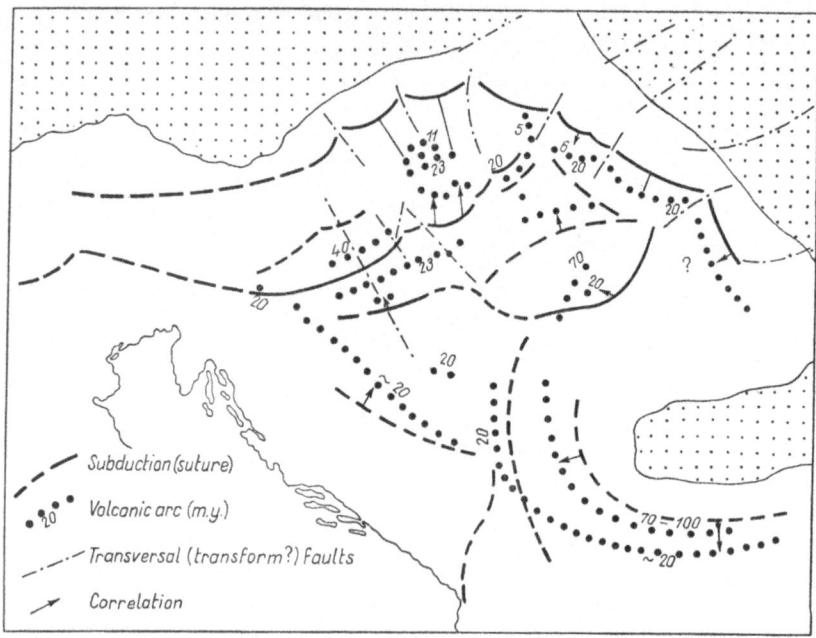

Fig. 3. The suture lines of subductions and the related volcanic arcs in the Alp-Carpathian-Balkan and Dinarid system.

Two further covered sutures occur presumably along the Rába-line under the Little Hungarian Plain and along the covered Upper Cretaceous-Paleogene flysch belt under the Great Hungarian Plain, the latter dipping according to geomagnetic data (fig. 6; *Pozsgay*, 1966) steeply (about 80°) towards NW and N.

The subducted sediments are transformed within the belt of subduction into metamorphites, granitoides, andesites, and basalts with an end-rest of ultrabasites (fig. 4). Thus subduction zones are accompanied by correlated descendent volcanic arcs, i. e. partial melts of the subducted chiefly sedimentary rocks being younger only by several million years than the beginning of the subduction.[*] Closer to the suture line the andesitic — or in case of a steeper subduction the ignimbritic belt —, and farther off the basaltic and alkaline volcanoes occur. In case of steep subductions the basalts are slightly alkaline (e. g. only basalts and basanites). The alkalinic-basaltic volcanism appears mainly along structurally weakened lines according to the decreased eruptive energy of their magma.

[*] *Stille* has already supposed such a correlation between the sedimentary sequence and the volcanic arcs of the Carpathians as mentioned kindly by Professor Dr. W. E. *Petrascheck*.

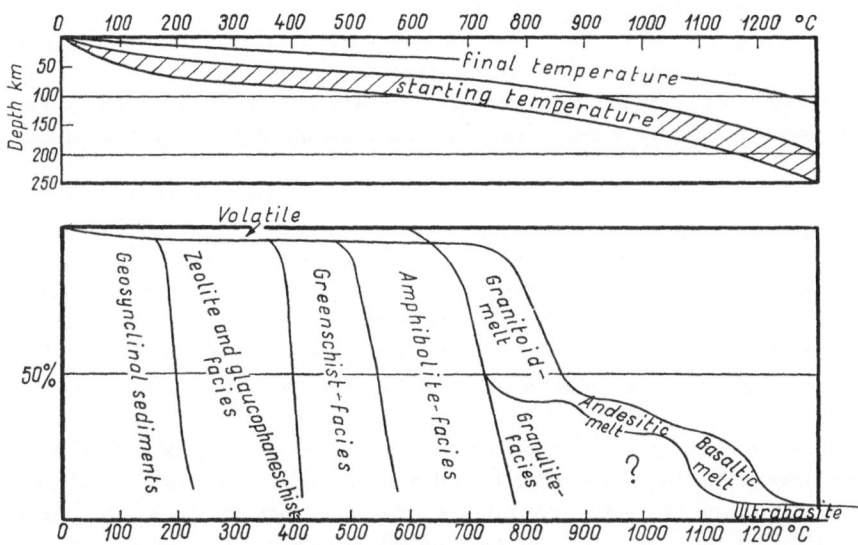

Fig. 4. Products of the subducted geosynclinal sediments as a function of temperature and depth.

In young volcanic areas, e. g. in the mainly Neogene subduction and volcanic system of the Carpathians-Pannonians the volcanic bodies are still mighty. In the high mountains of the Alps subducted earlier (mainly in the Cretaceous) most andesites are eroded. However, andesitic necks are known, e. g. in the Bacher Mts. and tonalite porphyries along the Judicaria (kind personal communication of Prof. Dr. E. *Clar* and Prof. Dr. *Exner*). In the deeper areas the younger basaltic derivatives of the Alp-subduction (E. g. Auvergne, W-Germany, Czechoslovakia, Silesia) are better conserved. Along the deep structural lines some volcanoes may occur also at the opposite side of the suture line (e. g. Euganeen).

The correlated volcanic belt of the Carpathian Klippen-belt subduction being garland-like dissected into seven sections (fig. 3) is represented by the Intra-Carpathian andesitic volcanic arc separated also into six or seven sections. Derivatives of subductions of steeper dip and faster rate form mixed ignimbrites (covered mostly by the Neogene of the Great Hungarian Plain).

On the basis of the K_2O/SiO_2 ratio of the volcanic rock (according to *Dickinson, Hatherton, Ninkovitch, Hays* and others) the depth of the primary "magma-chamber" was calculated* and proved to be 120 to 160 km for the andesitic ones,

---

* Even a method for the determination of the mass, thickness and breadth of the subducted sediments, including the differential velocities of the subduction from the mass and age relations of the andesites is elaborated *(Author, 1773 b)*.

150 to 250 km for the ignimbritic ones and 150 to 260 km for basalts and basanites. Some comparative calculations for other areas resulted similar values. These
depths of volcanic magma chambers correspond to those of the low velocity layer
of the Gutenberg channel characterized by a portion of partial melt, according to
the generally accepted views. These depths are presumably the most suitable for
partial melting.

Ignimbrites characterized by greater and variegated depths of magma chamber
and by steeper dip of subduction (fig. 2 and fig. 6) occur in the Pannonian Basin
between two parallel and synchronous suture lines of subduction. In this case the
drift and subduction of the posterior plate are furthered by the parallel movement of the first one above it. The sedimentary cover of the posterior plate is
subducted faster, it remains colder and reaches greater depths without melting,
then it produces violent ignimbritic eruptions characterized by high vapor contents.

In the front of the subduction mainly at the concave side of curved suture
lines there is usually an elevation of imbricate ("wedge") structure (fig. 7), e. g.
Central Mts. of Hungary, Slovakian Carpathians. Behind the suture line —
especially on the convex side of the inflexion — depressions occur, e. g. Little
and Great Hungarian Plain, (fig. 5).

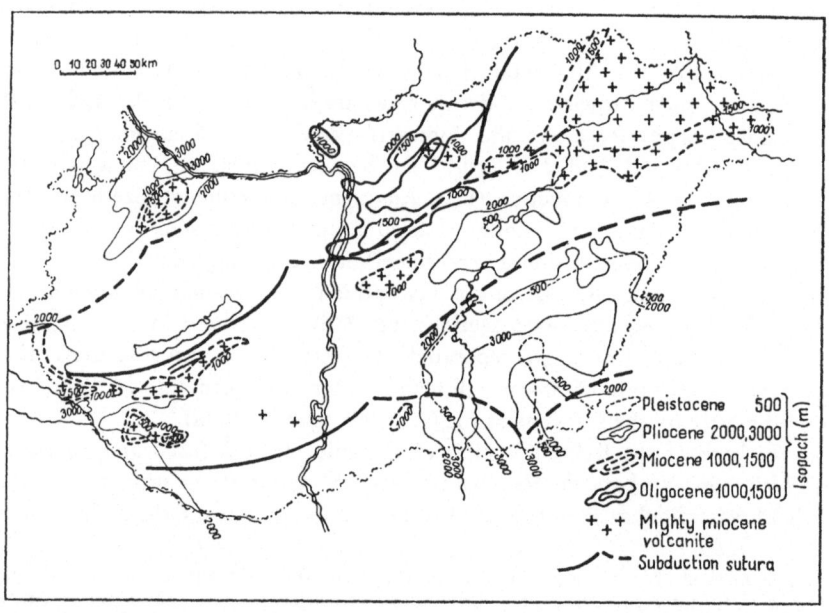

Fig. 5. Correlation between subduction and subsidence in the Pannonian Basin.

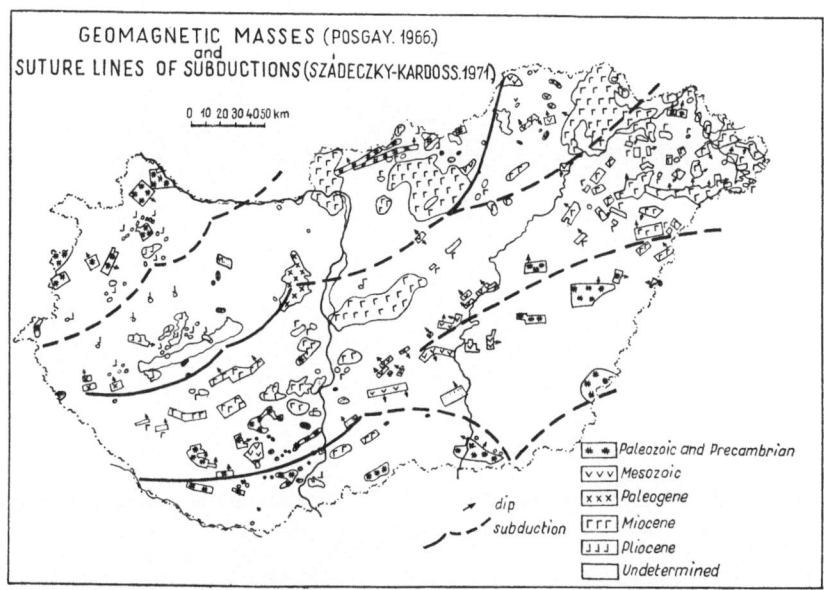

Fig. 6. Correlation between subductions and geomagnetic masses.

In the continental crust, the depression behind the subduction produces a partial transformation of the lower basaltic crust into eclogite facies representing seismically a sub-Moho "mantle" layer (fig. 8). The thin out of the crust of the Pannonian Basin and of other median masses may be partly due to this (*Stegena*), but relates to the higher position of the Gutenberg low velocity layer, too, according to the increased vapor pillow fed by more neighbouring subduction belts. The position of the basalt crust as partial melting product of the ultrabasic mantle and that of the low velocity layer containing also partial melt products are controlled primarily by temperature, thus their positions change as a rule similarly.

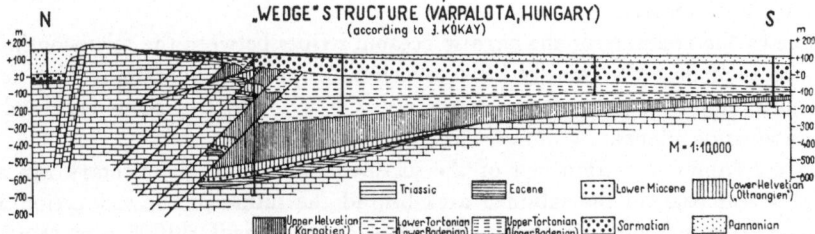

Fig. 7. Profile across the marginal subduction of the Hungarian Central Mts., Várpalota north of the Balaton lake, according to J. *Kokay*.

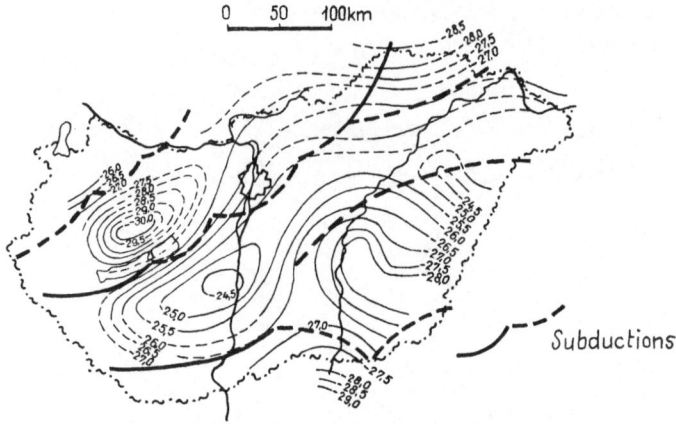

Fig. 8. Depth of the Moho-layer (km) and the positions of the suture lines of subduction in the Pannonian Basin.

2. Ore formations and distribution of trace elements

The oblique belt of subduction and the vertical channels of their volcanic derivatives represent the main transport ways of the ore forming solutions and melts.
Thus, the appearence of mineral resources near the surface are mostly connected
to these two main transport lines.

Along subduction zones ore forming volatiles are transported mainly as ascending and relatively low temperature aquaeous solutions and gases. The main
mineral resources correlated to the suture lines are therefore hydrocarbons and
metasomatic hydrothermal Fe and Pb-Zn ores (fig. 2). In the vertical igneous
channels the ore forming volatiles are connected to ascending magma and to
mostly high temperature solutions forming e. g. Au-Ag-Pb-Zn-Cu ores and skarns
connected to the subsequent andesites and late-kinematic granitoids. The ignimbritic volcanism is nearly ore-free, the ore-containing fluids escape in this case
into the atmosphere.

Due to the reclosing of the narrow oceanic stripes between the microcontinents
in median mass older, mostly "eugeosynclinal" ophiolites containing high temperature "liquid magmatic" ores (Cr, Ti-V, Cu, Ni, Co) also occur near and in
the subduction planes.

In the frontal elevation belt of the subductions bauxite, sedimentary Mn and
Fe ores and coal, in the subsided area behind the suture of the subduction and
between two subduction sutures evaporites occur. Massif sulfide ores (Kuroko
type etc.) connected to the eugeosynlines in the front side of the sutures *(Guild,*
1972) are restricted to this type.

Petrologic changes in the belt of subduction are accompained by changes in the distribution of trace elements. During the *progressive metamorphism* of clayey sediments volatiles produced by the subduction gradually emigrate from these rocks, transporting certain parts of the less volatile elements, too. Since these solutions pass through the whole series of the transforming rocks, the quantities of the element departing upwards and being supplied from the depth are compensating each other, but the quantity of most of the elements suddenly decreases at the lowermost ultrabasic rest of the series (Table I). This approximate dynamic equilibrium of the non volatile elements is expressed by the "isochemical transformation" of metamorphites.

Table I

Distribution of minor elements in the main rocks

	Maximum content	Minimum content
Clay	Li, Cs; Al, La, U; Mo, W Hg; Ga, Ge, Sn, Pb; As, Sb; S, Se F, J; B; C	
Carbonate rocks	Ca, Sr, Ba; Br	Na, Al, Sc; Si, Zr, Th; V;
Sand	Si; Y; Zr	Fe, Co; Cu, Ag; Cd; Ge, Sn Sc, U, Nb; Ni; Zn, Cd; — Cl, P
Granite	Na, K, Rb, Cs; Be, Ba; Y; Si, Zr, Th; Ta, Nb; Ga, In, Tl; Pb, Bi; — Cl, F; Ni	Mg, Ca; Cr, Mn, Bi, Se, J
Basalt	Al, Sc; Ti; V; Mn Cu, Ag; Zn; — P	Pt; Se
Ultrabasite	Mg, Cr, Fe, Co, Ni, Au, Pt	Li, K, Rb, Cs; Be, Sr, Ba; Y, La; Ti, Ta; Mo, W Cd, Hg; Ga, In, Tl; Sn, Pb; As, Bi; S, Te; F, Br, J; B; C; N

First the heavy sidero- and chalcophile rare elements bound incongruently in the rock-forming minerals are mobilized by the high pressure vapors formed during the metamorphism and partial meltings. During the crystallization of the melts they become camouflaged by the main elements of the new magmatites according to their ionic radii. In the course of the hydrothermal process they — being bound incongruently again -- are partly released and concentrated forming independent ore deposits.

Consequently, mainly the trace elements in a given magmatite are relatively concentrated, the ionic radii of which are commensurable to those of the main elements of the magmatite. The same trace elements form also the main ore deposits in the surroundings of the magmatic suite (fig. 9). According to this model the petrometallogenetic relations described by *Bilibin, Smirnov, Borchert, Abdulaev* and others originate in this way.

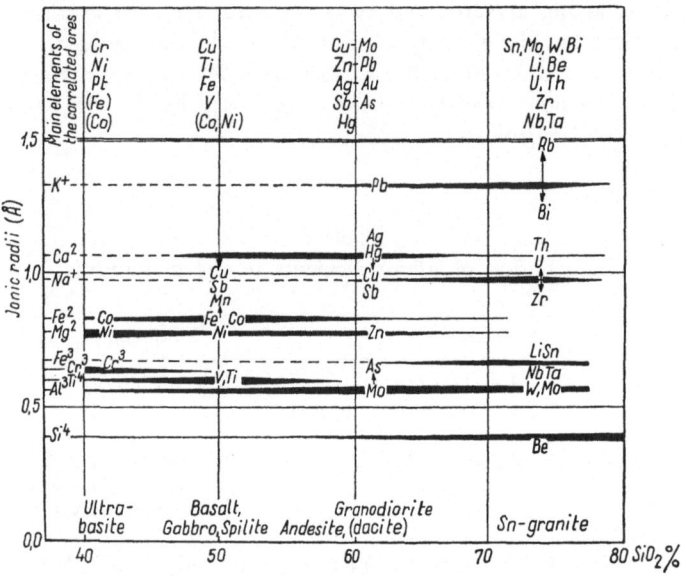

Fig. 9. Correlation between major and camouflaged main minor elements in the magmatic rocks.

The distribution of the trace elements depends mainly on their mobilities. In the belt if subduction and in the volcanic channels differences exist among the average litospheric melt, solution and gas-mobilities. These mobilities can be determined for different elements by geochemical-statistics. The melt mobility of an element (M_m) can be expressed by the ratio of its average concentration in the final granitic (C_g) melt and in the "molten" crust (C_m). Since $M_m = C_g / C_m$, and the sequence of melt mobility is essentially identical with that of "granitophily". The calculation summarized in Table II and fig. 10 shows that the granitophile elements, i. e. highly mobile in magmatic melts, are first of all the lithophile ones in the broader *(Goldschmidt)* sense, further three chalcophile elements of changing valency (Sn, Pb, Bi) as well as the Cl and F. *De Wijs* (1960) qualified the same elements as granitophile, only Ca, Ti and V was qualified to granitophile instead of granitophob. The term granitophob means actually the terms basaltophile- and ultrabasaltophile together.

78

Table II

Element	Concentration (p.p.m.) in the granites	clays	"molten crust"	Melt mobility (= 1. : 3.)	Solution mobility (= 2. : 3.)	Complex mobility ($\frac{= 4. + 5.}{2}$)
	1.	2.	3.	4.	5.	6.
Ag	0,05	0,1	0,07	0,71	1,43	1,07
Al	77,000	104,500	80,533	0,95	1,29	1,12
As	1,5	6,6	1,67	0,89	3,95	2,42
Au	0,0045	0,001	0,004	0,65	0,68	0,66
B	15	100	11,67	1,29	8,57	4,93
Ba	830	800	653,33	1,27	1,23	1,25
Be	5,5	3	3,8	1,45	0,79	1,12
Bi	0,01	0,01	0,009	1,11	1,10	1,10
Br	1,7	6	2,13	0,80	2,82	1,81
C	0,03	2,28	0,023	1,30	99	50,15
			(0,70)	(0,04)	(3,25)	(2,27)
Ca	15,800	25,300	32,933	0,47	0,77	0,62
Cd	0,1	0,3	0,13	0,77	2,31	1,54
Ce	100	50	68,17	1,46	0,73	1,09
Cl	240	160	176,67	1,36	0,91	1,13
Co	5	20	18,33	0,27	1,09	0,68
Cr	25	100	83,33	0,30	1,20	0,75
Cs	5	12	3,67	1,25	3,27	2,26
Cu	20	57	47	0,43	1,21	0,82
F	800	500	656,67	1,22	0,76	0,99
Fe	27,000	33,300	46,533	0,58	0,72	0,65
Ga	20	30	19,33	1,03	1,55	0,79
Ge	1,4	2	1,43	0,97	1,39	1,18
Hg	0,08	0,4	0,083	0,96	4,82	2,89
H_2O	0,61	4,76	0,59	1,03	4,60	2,81
In	0,26	0,05	0,25	1,04	0,20	0,62
J	0,4	1	0,4	1,0	2,5	1,72
K	33,400	22,800	25,033	1,33	0,91	1,12
La	60	40	49	1,22	0,82	1,02
Li	40	60	31,67	1,25	1,89	1,57
Mg	5,600	13,400	18,733	0,29	0,72	0,50
Mn	600	670	1,066	0,56	0,62	0,59
Mo	1	2	1,13	0,88	1,77	1,32
N	20	600	19,3	1,04	31,1	16,07
Na	27,700	6,600	24,933	1,11	0,26	0,68
Nb	20	20	20	1,0	1,0	1,0
Ni	8	95	58,66	0,13	1,62	0,87
P	700	770	933,33	0,75	0,83	0,79
Pb	20	20	16	1,25	1,25	1,25
Pd	0,01	0,013	0,76	—	—	—
Rb	200	200	148,33	1,35	1,35	1,35
S	400	3,000	366,67	1,08	8,19	4,63
Sb	0,2	1,5	0,2	1,0	7,5	2,21
Sc	3	10	10	0,30	1,00	0,65

Element	Concentration (p.p.m.) in the granites	clays	"molten crust"	Melt mobility $(= 1. : 3.)$	Solution mobility $(= 2. : 3.)$	Complex mobility $\left(= \dfrac{4. + 5.}{2}\right)$
	1.	2.	3.	4.	5.	6.
Se	0,05	0,6	0,05	1,00	1,20	6,5
Si	323,000	238,000	295,333	1,09	0,80	0,94
Sn	3	10	2,5	1,20	4,00	2,60
		(6)			(2,2)	
Sr	300	450	346,67	0,86	1,29	1,07
Ta	3,5	3,5	2,49	1,40	1,41	1,40
Th	18	11	13,00	1,38	0,85	1,12
Ti	2,300	4,500	4,533	0,50	0,99	0,74
Tl	1,5	1	1,07	1,40	0,96	1,18
U	3,5	3,2	2,50	1,40	1,28	1,34
V	40	130	93	0,43	1,39	0,91
W	1,5	2	1,33	1,12	1,50	1,31
Zn	60	80	83,33	0,72	0,96	0,84
Zr	200	200	166,67	1,19	1,20	1,19
Y	34	30	29,33	1,16	1,02	1,09

Since diluted solutions of the crust together with the elements dissolved in them originate mainly from the dehydration of the clays and metaclays, the solution mobility (M_s) is determined by the weight-per cent ratio of the average concentration of the element in the clays (C_c) and in the "molten" crust. Hence, $M_s = C_c / C_m$ and the sequence of solution mobilities is essentially that of "argillophily". First of all anion forming sedimentophile and chalcophile elements are argillophile, e. g. N, B, C, halogens, the sulphur- and arsenic group and alkali elements. Thus, the separation of the volatiles from the non-volatiles takes place mostly in the solution phase. The extension of the term "volatile" over the melts is a rather arbitrary extrapolation.

To perform this calculations the data of *Vinogradov* (1962), and of *Turekian* and *Wedepohl* (1960) were used. According to *Vinogradov* the "molten" crust can be approximated by the mixture of two parts of granite and one part of basalt. The composition of the real complex crust determined by *Ronov* and *Yaroshevski* shows somewhat greater contents of (ultra)basic elements, e. g. Fe, Mg, Mn, Ti and Ca (Table III) as the real complex crust contains ultrabasic erosion products, too, and the subduction of the calcium-rocks, limestone and other sedimentary carbonates forming great rigid masses is often hindered *(Author, 1971).*

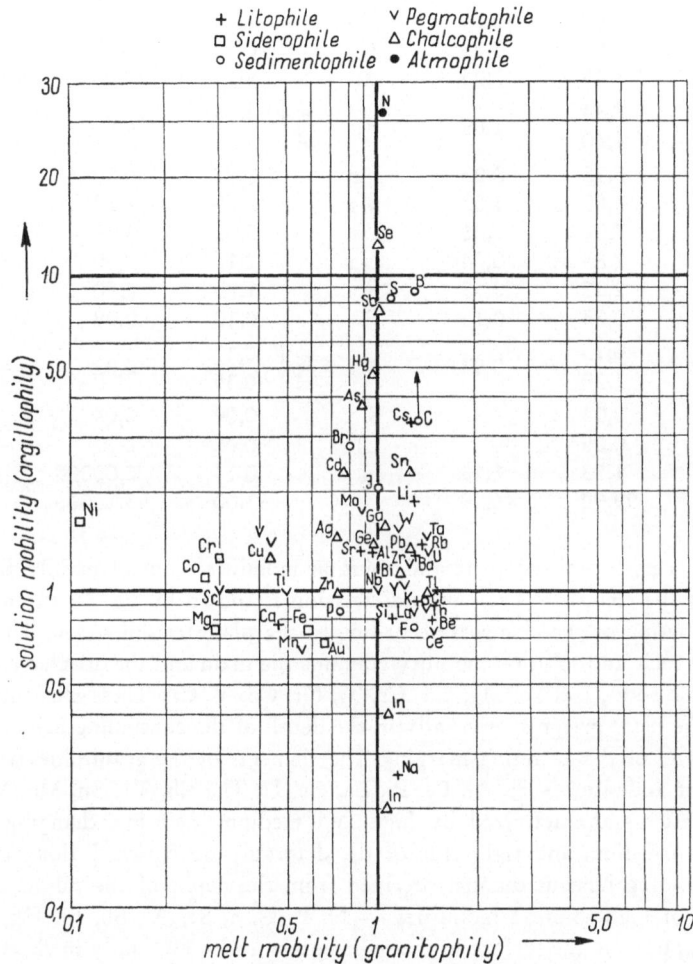

Fig. 10. Geochemical characters of the elements in function of their melt mobilities and solution mobilities.

Table III
Composition of the complex and molten crust

Oxides	Complex crust % (Ronov-Yaroshewski)		Cations		Primary molten crust %	Difference between molten and complex crust %
		Oxygen				
SiO_2	57,64	30,74	Si	26,90	29,53	+2,63
TiO_2	0,88	0,35	Ti	0,53	0,45	—0,08
Al_2O_3	15,45	7,25	Al	8,20	8,05	—0,15
Fe_2O_3	2,43	1,69	Fe 5,04 {	1,69	4,65	—0,39
FeO	4,30			3,35		
MnO	0,15	0,03	Mn	0,12	0,10	—0,02
MgO	3,87	1,53	Mg	2,34	1,87	—0,47
CaO	7,01	2,01	Ca	5,00	3,29	—1,71
Na_2O	2,87	0,74	Na	2,13	2,49	+0,36
K_2O	2,32	0,39	K	1,93	2,50	+0,57
P_2O_5	0,23	0,13	P	0,10	0,09	+0,01
C_{org}	0,13	0,94	C 0,48 {	0,13	0,02	—0,46
CO_2	1,29			0,35		
S	0,04	—	S	0,04	0,04	±0,0
Cl	0,05	—	Cl	0,05	0,02	—0,03
H_2O	1,33	1,19	H	0,14	0,00006	—0,13994
Σ	99,99	46,99		53,00	53,10006	

In the crust litospheric melt and solution mobilities control mainly the distribution of the elements. Their mutual relation is shown by fig. 10 demonstrating the distribution of the elements according to the plate-tectonics, too. The field of minimum melt and solution mobility contains the main and the ore elements of the (ultra)basic rocks, i. e. Fe, Mg, Ca, Cr, Ti, Ni, Co, V, Cu. These elements appear next to the plate margins, especially in the depth of the consuming belts along the trenches and of the accreting margins. The elements of the granitoides and of the subsequent andesites — Si, Al, Be, Li, La, Ce, U, Th, Nb, Ta; Sn, Mo, W, Cl, F, Zn, Pb, Ag — characterized by high and medium melt but changing solution mobilities (medium and right side of the diagram) are enriched along the vertically ascending igneous channels derived from the depht of the subduction belt. Elements of high solution mobilities — N, B, C, S, Se, As, Sb, Bi, Hg, Cs, Na, Br, J — appear on the surface in the sedimentary rocks, especially in clays.

The main elements of pegmatitic-pneumatolitic mineralization are of greatest melt mobility, thus they are mostly ascendent and magmatic (fig. 11). According to the values of the melt and solution mobilities hydrothermal ores are controlled by combined magmatic (melt) and solution processes. The telescoping succession of *Emmons-Berg-Schneiderhöhn* for hydrothermal ores in veins corresponds essentially to a complex mobility sequence computed as mean values of the lito-

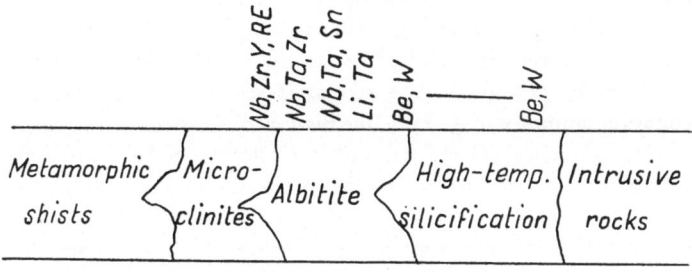

Fig. 11. Generalized sequence of metasomatic rocks and ores in the apical parts of granitoid bodies, according to data of A. A. *Beus*, 1970.

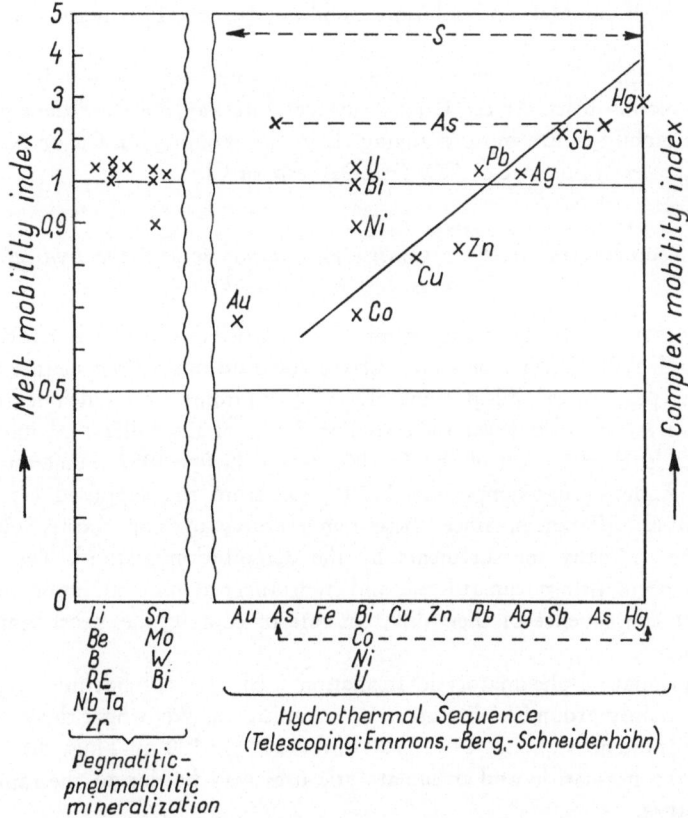

Fig. 12. Sequence of the postmagmatic ore formations in function of the complex mobilities.

spheric melt and solution mobilities (fig. 12). According to their different valencies the anion and cation forming elements, sulphur and arsenic, occur several times in these sequences, at lower places of the mobility than expressed by their total complex mobility indices. The schematic sequence of the horizontal metallogenic zonation supposed e. g. by *Gabelman* and *Krusiewski* (1972) is similar to the telescoping sequence. Therefore, such zonations can be characterized by the complex mobility indices progressing from the melt to the solution mobilities, too.

In the course of metamorphism the distribution of the elements, determined in a first approximation on the basis of several characteristic metamorphic and metasomatic parageneses by *Korshinski* for three temperature ranges, seems to be controlled by the transition of the melt and solution mobilities, i. e. approximately by the complex mobility sequence, as well as the distribution of elements observed e. g. by *Shaw* in the case of the progressive metamorphism of New Hampshire. The quantities of elements of small mobilities (Ni, Cu) decrease, those of high complex and solution mobilities (Li, Sr, Pb) increase towards the depth.

These litospheric solution mobilities control solutions of the volcanic channels and subduction belts. At the Earth's surface more varying conditions prevail and solution mobilities are more changing. E. g. the mobility of Ca or Na considerably increases in presence of CO_2 (HCO_3) resp. of Cl.

3. Geochemistry of the pegmatite-pneumatolytic and the hydrothermal ores and related rocks

The conspiciously different chemical pecularities of the pegmatitic-pneumatolitic and hydrothermal ores and related rocks do not follow neither from their plutonic (pegmatites and pneumatolytes) or plutonic and volcanic (hydrothermalites) relations, nor from their similar depth of crystallization mostly in the uppermost level of six km of the Earth's crust, *(Author* 1957, numerous papers of the St. Andrews ore-symposium 1970), nor from the supposed crystallization temperature differences, since these temperatures overlap mostly one another according to many measurements by the derepitation method. The differences between pegmatitic-pneumatolytic and hydrothermal mineralization seem rather to be the consequence of their different rate of cooling, i. e. geothermal gradient (fig. 13).

The pegmatitic-pneumatolytic formation with tin-molybdenum-tungsten ores appears mainly around highly ascending granitic cupolas where the value of geothermal gradient is about 40^0 C / 100 m to 130^0 C / 100 m. Thus, the crystallization of the pegmatitic and pneumatolytic ores may be due to the rapid drop of temperature.

In the contrary, the temperature gradient of the hydrothermal ore formation mostly does not exceed the value of 40^0 C / 100 m. This is, however, obviously

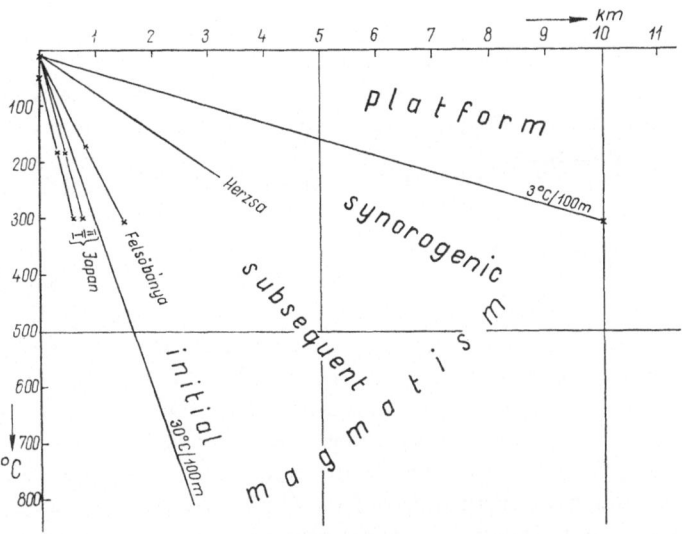

Fig. 13. Changes of the geothermal gradients in the crust of the earth.

insufficient to produce crystallization from the dilute hydrothermal solutions even in the case when the solubility of the metallic ions is increased by the presence of alkali-halogenides. As indicated by the calculated different solubility values hydrothermal ores represent rather chemical precipitates of the ascending solutions relatively rich in metallic heavy cations and of the descending or quasi-lateral-secretional solution relatively rich in anions of the sulphur and arsenic group. In this way hydrohtermal ore minerals may form an even almost closed vein system surrounded by clays (metaclays) when they become compacted and dehydrated by hot solutions ascending from the environment of a magma body.

The rhythmic alternation of the hydrothermal vein minerals supports this model. In the course of uniform dilatation of the vein fissure the barren minerals (quartz, carbonates etc.) crystallize relatively rapidly tightening more and more the fissure. When the fracture is almost closed, the ore minerals are precipitated by the interaction of the slower infiltrating quasi-descendent sulphidic and the ascendent heavy-metallic solutions. The further uniform dilatation opens the fissure again and the crystallization of the subsequent rhythm follows.

Hence, most hydrothermal ores are supposed to represent diplogenetic products of the quasi-descendent argilogenic anions and of the ascendent magmatogenic heavy-cations.

On the contrary, both the heavy cation and the (oxydic) anion of the pegmatites and pneumatolytes are of igneous origin, i. e. they represent mostly monogenetic formations.

According to this model the main condition of the hydrothermal ore formation is a possibility of exchange between the igneous mass and the level of clays or

85

metaclays — either directly or by fracture lines. Geological evidences of this original situation, however, often vanish, the clayey roof may be metamorphosed or eroded and only the partly descendent ore vein of the underlying crystaline basement remains. A subsequent recrystallization of the ore may even contraindicate the original situation.

However, there exist other indirect evidences of this new model. According to this model the composition of the country rock may determinate decisively the type of ore. Especially the effect of the changing sulphur content is remarkable. Pegmatitic-pneumatolytic ores are mainly oxidic, according to the anion deriving from the magma poor in sulphur. They occur mostly at the top of highly differentiated, strongly ascendent ignous masses crystallizing in areas of older igneous or strongly metamorphized and carbonated rocks. — The average sulphur content (counted as SO_3) of the continental clays and carbonate rocks is 0,43 resp. 2,36 per cent (Table IV) according to *Ronov* and *Yarosewski*. On the contrary, it is 0,02 to 0,03 per cent in the oceanic sediments and volcanic rocks. Accordingly, the hydrothermal ores of the continental areas are mostly sulphidic, while those of the eugeosynclines at the oceanic margins are rather oxidic (Lahn-Dill type) and only partly sulfidic (e. g. massif sulphide deposits, Kuroko type).

Table IV

Sulphur content of rocks (according to *Ronov* and *Yarosewski* 1969)

	Continental	Geosyncline SO_3-content	Oceanic
Sands	0,29	0,12	—
Clays	0,43	0,11	—
Carbonate rocks	2,36	0,03	—
Volcanites	0,03	0,03	—
Total S	0,04	—	0,02

Greisens and albite or microcline belts in connection with pegmatitic-pneumatolytic mineralization (fig. 11) are essentially of pure igneous origin, too. — On the contrary, propylitization preceding and following the hydrothermal ore formation, represents H_2O — H_2S — CO_2 metasomatism due to the volatiles deriving from sedimentary rocks. The propylitic chlorite-sericite-carbonate alteration may be accompanied by further metasomatic mineral assemblages, mostly in the following sequence from below upwards: (propylite), kaolinite-sericite-quartz, pyrophyllite-dickite-quartz, diaspore-quartz, pure quartz. This sequence of metasomatic alteration may be several hundred metres thick *(Drovenik, Radonova and Velinov, Skripcenko, Naboko, Romanova & Petrachenko, Takeushi & Abe).* In the Sredno-Gora series of Bulgaria and Yugoslavia this sequence is e. g. about 700 metres thick. The silica zone occuring often in the upper- and lowermost level derives from the surplus of silica released by clay mineralization and chloritization of the primary igneous feldspars and pyroxenes or amphiboles.

These metasomatic series were formed mostly by the sulphuric acid derived from the oxydation of the sedimentary pyritic deposits accompanying them. In other cases sulphuric acid originates from ascendent endogenic H_2S being often an additive factor to the predominant sedimentary process, too.

The endogenic ascendent factor becomes predominant in the areas of intense late-volcanic activity, e. g. in the NE of the Carpathian Basins, where sedimentary rocks are accumulated in the depth by subductions from several directions. In this case a strong potash-metasomatism develops in connection with the hydrothermal ore formation due to the late-igneous alkali-rich solutions. The potash-metasomatics, which can be followed downward farthest in the surface spreads with a mushroom-shape according to the near-surface mobility of these solutions. Potash-metasomatism may form a metamorphic depth-sequence of several mineral assemblages of the zeolite and greenschist facies: chlorite-carbonate-adularia-zeolite assemblage in the uppermost, epidote-chlorite-albite assemblage in the middle and actinolite-epidote-albite assemblage in the lowermost level (*Széky-Fux* 1965, *Vasilevski* 1970).

In accordance with the alkaline reaction of the potash solutions the potash-metasomatites show a high oxydation degree with a considerably increasing ratio of Fe_2O_3/FeO. The rich gold-silver ores — exhausted in the Charpathian area mostly during the preceding centuries — appear mainly in these oxydized rocks, separatedly from the wide lower level of the normal polymetallic ore formations.

The formation of potash-metasomatites is analogous to the pegmatitic-pneumatolytic microlinization. Sometimes the sodium-metasomatites, e. g. sodium-rhyolite, corresponding to the pegmatitic-pneumatolytic albitization also occur in connection with the hydrothermal ore formation.

References

Dewey, J. P. & J. M. *Bird* (1970): Journ. Geophys. Res. *75*, 2625.
de Wijs, H. J. (1960): Geol. Mijnbouw *22*, 57—63.
Dickinson, W. R. & T. *Hatherton* (1968): Science, V. 157, 801.
Drovenik, M. (1960): See Radonova, T. G. and I. Velinov in Problems of hydrothermal ore deposition, St. Andrews, 1967, p. 369.
Gabelman, J. W. & S. V. *Krusiewski* (1972): XXIV. Internat. Geol. Congr. Montreal *4*, 88.
Korzhinsky, D. S. (1950): Internat. Geol. Congr. 1948. London, Proc. *3*, 65—72.
Lapadu-Hargues, P. (1954): Bull. Soc. geol. France, Ser. 5, *13*.
Mackay, R. A. (1946): Econ. Geol. *41*, 13—46.
Mackay, R. A. (1950): Internat. Geol. Congr. 1948, *12*, 165—191.
Naboko, S. I. (1963): Acad. Sci. USSR (Russ).
Ninkovich, D. & J. P. *Hays* (1969): Acta Int. Cong. volc. Thera, Lamont-Doherty Obs.
Ovchinnikov, L. N. (1956): Doklady AN SSSR *1*.
Ovchinnikov, L. N. (1956): Trudy IGEM AN SSSR *6*.
Radonova, T. G. & I. *Velinov* (1967): Probl. of hydroth. ore dep. St. Andrews, 368—372.
Romanova, R. I. & E. D. *Petrachenko* (1967): Probl. of hydroth. ore dep. St. Andrews, 373—375.

Ronov, A. B. & A. A. *Yarosevszki* (1969): The Earth's crust and Upper Mantle *13*, 37.
Shaw, D. M. (1956): Bull. Geol. Soc. Amer. 67, 919—934.
Skirpchenko, N. S. (1967): Probl. of hydroth. ore dep. St. Andrews, 375.
Stegena, L. (1971): Anyag- és Energiaaramlási Ankét, Budapest, Akadémiai Kiadó, 199.
Stille, H. (1953): Beitr. Geol. Jahrb. *8*, Hannover.
Szádeczky-Kardoss, E. (1958): Acta Geol. Acad. Sci. Hung. *5*, 3—4.
Szádeczky-Kardoss, E. (1971): Geonómia és Bányászat, 83.
Szádeczky-Kardoss, E. (1973): Földtani Közlöny (Geol. Mitt.), (in print).
Szádeczky-Kardoss, E. (1973): Földtani Kutatás (Geol. Investigations), (in print).
Széky-Fux, V. (1965): Acta Geol. Acad. Sci. Hung. 259.
Széky-Fux, V. (1965): Die Erde, Jena, 104—197.
Széky-Fux, V. (1970): Telkibánya ércesedése és kárpáti kapcsolatai (The Telkibánya
  mineralization and its Inter-Carpathian connexions), Akadémiai Kiadó, Budapest.
Takeuchi, T. & H. *Abe* (1967): Probl. of hydroth. ore dep. St. Andrews, 381.
Turekian, K. K. & K. H. *Wedepohl* (1961): Bull. geol. soc. Amer. 72, 2.
Vassilievskii, M. M. (1967): Probl. of hydroth. ore dep. St. Andrews, 384.
Vinogradov, A. P. (1962): Geokhimiya, 555—571.

Mineral Distribution and Geological Features of the Philippines

Gabriel *Santos*, Jr.*

Abstract

The Philippine deposits are divided into two main groups based on their locations in the arc-trench system, namely frontal arc and third arc. Deposits occurring in the frontal arc are characterized mainly by the predominance of nickel, laterite, chromite and cupriferous massive sulfide, under a geologic setting distinguished by ultrabasic and basic volcanics and subordinate intermediate intrusive-metamorphic (schist) rocks. These regional rocks are adjacent to deep troughs or trenches.

Deposits belonging to the third arc are primarily characterized by the gold-bearing copper sulfides with pyrite and minor magnetite and molybdenite. The mineralizations are closely related to intermediate intrusives and intermediate-basic volcanics with occassional ultrabasics.

Southeastern Luzon could probably exhibit the features of both the frontal and third arcs. It is suggested that the Sulu Sea Basin could probably have been formed as a result of the extensional rifting of a part of the frontal arc.

Zusammenfassung

„Mineralverteilung und geologische Merkmale der Philippinen"

Auf Grund ihrer Lage im Bogen-Graben-System lassen sich die Lagerstätten der Philippinen in zwei Hauptgruppen teilen, nämlich die des frontalen und des dritten Bogens. Die Lagerstätten des frontalen Bogens sind vor allem durch das Vorherrschen von Laterit, Chromit und massigen, kupferhältigen Sulfiden mit begleitender Zinkblende gekennzeichnet. Die geologischen Formationen, in denen

---

* G. *Santos*, Philippine Atomic Research Center, Diliman/Quezon City, Rep. of the Philippines.

sie sich befinden, zeichnen sich durch die Anwesenheit von ultrabasischen bis intermediären Vulkaniten sowie untergeordnet von intermediären intrusiv-meta-morphen (verschieferten) Gesteinen aus. Diese Gesteine liegen tiefen Trögen oder Gräben benachbart und werden in den südöstlichen Philippinen von der Philip-pinen-Störung abgeschnitten.

Die Lagerstätten des dritten Bogens sind vor allem durch die goldführenden Kupfersulfide mit untergeordnetem Pyrit, Magnetit und Molybdänit gekennzeich-net und eng mit intermediären Intrusivgesteinen und Vulkaniten, fallweise auch kleineren ultrabasischen Körpern, verbunden. Im südöstlichen Luzon sind wahr-scheinlich Merkmale sowohl des frontalen wie des dritten Bogens vorhanden. Es wird angenommen, daß das Becken der Sulusee wahrscheinlich durch ausgedehntes Rifting eines Teils des frontalen Bogens entstanden ist.

Introduction

The purpose of this paper is to relate the mineral deposit distribution of the Philippines and the geologic features in the light of its island arc structures. This work is a part of a research program of the Philippine Atomic Energy Commis-sion investigating the application of trace element distribution in mineral sul-fides to the study of metallogenic provinces. Based on available data on Philip-pine mineral deposits and personal fieldworks, this investigation classifies mineral deposits on the basis of their geologic setting. The setting, in this case, is visualized in terms of elements of an arc-trench system (Karig, 1970). In the course of this study, attempts were made to explain certain "enigmatic" geologic features of the Palawan-Sulu Sea area.

General Setting

The Philippine Islands have a roughly triangular areal form and are located in the western Pacific between latitudes 5^0 and 20^0 North and longitudes 115^0 and 127^0 East. These islands include Luzon, the largest island in the north; Mindoro, Masbate, Panay, Negros, Cebu, Bohol, Leyte, Samar and other smaller islands comprise the central Philippine group; Mindanao, the second largest island, and the Sulu group of islands constitute the southern portion of the Philippines and are separated from Palawan by the Sulu Sea. (see Fig. 1). The eastern part of the archipelago from Luzon through the central Philippines and Mindanao constitute the mobile belt while the Palawan and Sulu Sea areas are considered as the stable region *(Gervasio, 1964)*.

The areal distribution of principal oceanic troughs, mountain system and foci of deep-seated tectonic earthquakes and particular major mineral deposits have fairly good correspondence in the Philippine island arc system. Apparently, there seems to be some fundamental regional control governing metallization.

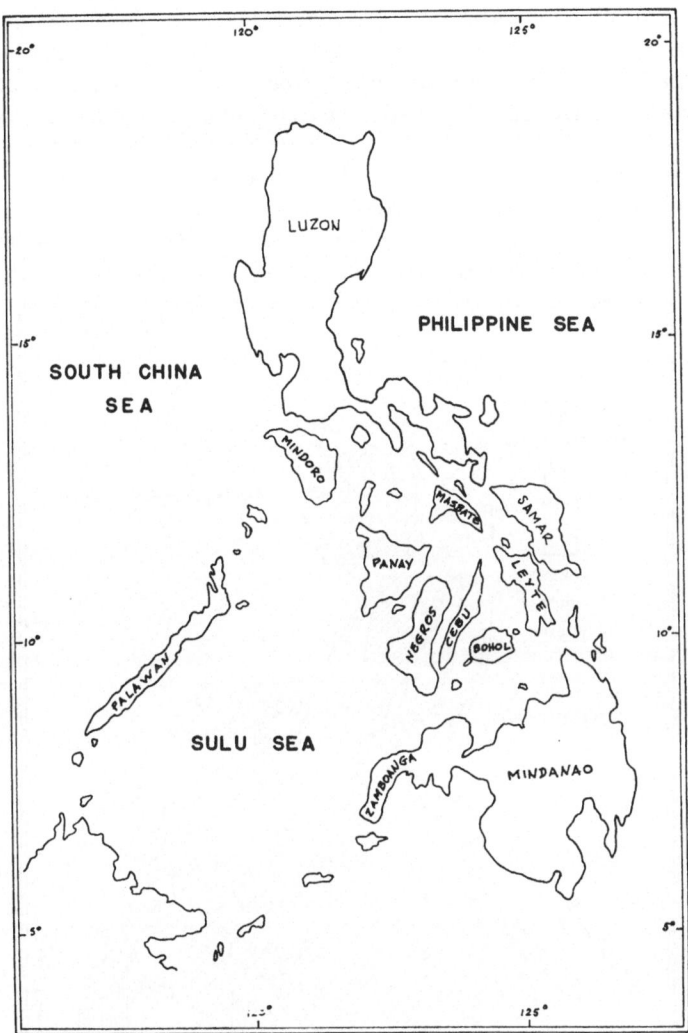

Fig. 1. Geographic Location of the Philippines.

The geology of the Philippines has been summarized in the "Geological Map of the Philippines" (1963), published by the Philippine Bureau of Mines. Other students of the Philippine geology include *Irving* (1949), *Corby* & al. (1951), *Ranneft* & al. (1960), *Gervasio* (1964) and *Melendres* (1971). The regional occurrences of mineral deposits particularly, copper-gold deposits were investigated by *Kinkel* & al. (1957), *Tupaz* (1960) and *Bryner* (1969).

91

Volcanic features dominate the regional geologic scene. Visualized in terms of the elements of an island arc system as defined by *Karig* (1970), the Philippine arc is roughly subdivided into parts of the frontal arc, inter-arc basin and third arc as shown in Figure 2. The location and dimensions of the basins were compiled by *Melendres* (1971). According to Karig, the third arc may be considered as a remnant of a former frontal arc and as a result of the extension between the frontal arc and the third arc, and inter-arc basin is formed.

In Fig. 2, the frontal arcs are situated in west Luzon-Palawan and southeastern portion of the archipelago. Parts of the third arc are found in northern Luzon

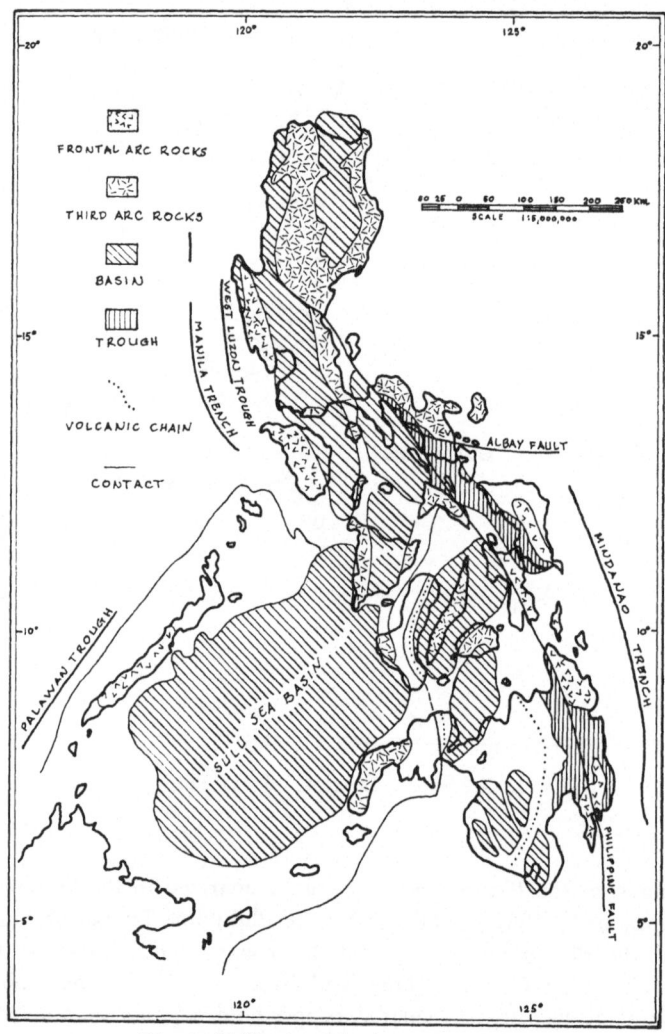

Fig. 2. Philippine Island arc System.

and in the islands of Central Philippines. The southeastern part of Luzon is classified as belonging more to the third arc than to the frontal arc. Elongate and narrow, locally discontinuous inter-arc basins separate the frontal arc from the third arc. The Sulu Sea basin, however, appears to approximate clearly the postulated origin by extensional rifting of inter-arc basin as observed by *Karig* (1970) in the Tonga-Kermadec, New Hebrides and Marianas island arc systems.

The frontal arcs are bounded by the Manila Trench and Palawan Trough in west Luzon-Mindanao and the Palawan island respectively; and the Mindanao Trench in eastern Samar, Leyte and Mindanao islands. The first marine geophysical surveys of west Luzon-Mindanao were conducted by *Ludwig* & al. (1967) and *Hayes* & *Ludwig* (1967) which led to the delineation of the Manila Trench and West Luzon Trough.

The localized small "basins" and "troughs" found in the archipelago are considered as parts of the inter-arc basin though they may not appear as much in the more ideal sense of the word as defined by Karig. This condition is brought about by the complexity of the Philippine arc to an arc system, wherein two oppositley adjacent facing zones of convergence are in evidence. Arc to arc relationships are possible according to *Isacks* & al. (1968). The subduction zones are delineated by the Manila Trench in western Luzon and Mindanao Trench in eastern Mindanao as shown in Figure 2. The Mindanao Trench system has the typical features of the classical arc-trench system as regards seismicity, gravity anomalies and volcanism. Associated seismicity of the Mindanao region is fairly high in comparison to the west Luzon region. The focal depths of earthquakes in Mindanao area exceeds 300 kilometers. In west Luzon, seismic activity is restricted to depths well below 300 km *(Hayes & Ludwig, 1967; Morante, 1970)*.

On the basis of dimension, the Mindanao Trench is much longer than the Manila Trench, which could account in part, for the latter's shallower seismic activity. *Datuin*, (1972) thinks that east of Luzon is probably the site of a developing trench on the basis of occurrences of earthquakes with shallow focal depths. He recognizes the presence of two blocks which represent the Philippine mobile belt, the Luzon and the Visayan-Mindanao blocks. The Philippine fault which passes through these blocks changes from dextral in Luzon to sinistral in the Visayan-Mindanao region. According to *Kintanar* (1966), "the Philippine fault is the only tectonic feature of regional consequences", though it is reasonable to expect that this fault has other branches. *(Irving, 1951; Melendres, 1971)*. An inferred transform fault is probably developing. The trace of the inferred fault could pass through the Mindoro Reentrant and the Albay Reentrant. The Albay fault is probably a part of this transform fault. The presence of this rift or fault was suggested recently by *Santos* & *Wainerdi* (1969). Thrust faults appear to be a common feature of the frontal arc.

According to *Kintanar* (1966) based on 400 stratigraphic sections in various areas, radiodating, depositional environment, petrology and regional data ... "the Philippines has been an archipelago since Upper Paleogene (Eocone-Oligocene). The position and shape of the islands have fluctuated with time, but the overall pattern has remained." The composite thickness of the (sedimentary) section is about 70,000 feet which are sub-divided into "five groups, each separated by an unconformity in areas of positive movement". The section "varies from mudstone to conglomerate and from limestone to chert." In general the sedimentary rocks are characterized as quartz-poor except in Palawan and southwestern Mindoro areas. Deposition environments ranges from abyssal to continental with common occurrences of turbidite deposits. Kintanar noted that since Eocene time the islands have been separated by deep troughs inter-connected by volcanic ridges on the sea floor. At times these ridges rise high enough to form islands and reefs. The Tertiary basins and troughs locations and boundaries compiled by *Melendres* (1971) are shown in Fig. 2.

Frontal Arc Rock Suite

Rocks found in the frontal arc are characterized by extensive ultrabasic- basic intrusive, of the Alpine type *(Thayer, 1960)* to basic volcanics with associated thick sedimentary rocks deposited in basins of miogeosynclinal type. The ultrabasic rocks consist mainly of serpentinized peridotite and dunite with minor gabbro and lesser pyroxenite. The rocks are typical of, though not confined to frontal arc environments. The volcanics are represented by the spilites-keratophyres and pillow lavas, most are interbedded with or intruded in marine sedimentary rocks. In the Mindoro-Palawan area, the sedimentaries consist of "arkosic sandstone, wacke and shale associated with quartzite" *(Gervasio, 1966)*.

A basement complex of pseudostratigraphic sequences called "ophiolite assemblages" as defined by *Gervasio* (1966) and *Dickinson* (1971) are overlain by Triassic (?) to Jurassic graywacke-shale-chert limestone in the Palawan-Mindoro and western Zamboanga areas. *Gervasio* (1966). The ophiolites consist partly of greenchist and basic amphibolite schist and gneiss with meta-graywacke, phyllite and slate.

The age of the basement complex which are found mostly in Palawan and Mindoro is probably Carboniferous to at least Permian based on discordant fossil-bearing sedimentary rocks. *Gervasio* (1966).

It is significant to note that the structural trend of the Triassic-Jurassic miogeosynclinal sedimentaries is north-south while the older folded belt of Palawan is northeast. *Gervasio* (1966; 1968). *Motegi* (1971) recognized this structural northeast "Palawan Trend" and suggests that this "ancient lineament" is closely related to the distribution of ore deposits. Both *Motegi* and *Ranneft* & al (1960) noted the strong northeast-trending tectonic lines in Zamboanga Peninsula.

On the basis of the juxtaposition of the north-south trending Tertiary quartz-rich sedimentary formations and the northeast trends of the pre-Tertiary igneous-sedimentary rocks of Palawan-Mindoro areas and in the light of new plate tectonics *(Dickinson, 1971; Dewey & Bird, 1970; Isacks & al., 1968)* it is suggested that rifting of the mio-geosynclinal rocks of Palawan and Mindoro occured probably from Carboniferous through Permian and that collision probably took place at the close of Jurassic to Early Cretaceous.

It may further be inferred that the Sulu Sea Basin was formed as a result of extensional rifting of the Palawan frontal arc on the basis of the following: (a) equivalent rock types found in Palawan and Zamboanga peninsula, (b) submarine morphology of Sulu Sea, *Irving* (1961), (c) the existence of the northeast "Palawan Trend" reflected by the Sulu island group through Zamboanga, (d) evidence of rapid subsidence of Sulu Sea, *Daleon* (1971).

Sediment thickness and distribution in the Sulu Sea may further substantiate this inference.

Third Arc Rock Suite

Intermediate intrusive-volcanics (diorite-andesite complexes) appear to be the distinctive rock of the third arc. Indeed, the core of structural highs in central Luzon consits of elongate synorogenic batholitic diorite masses which were intruded "near the end of middle Miocene time". These plutons generally trend northsouth *(Fernandez & Pulanco,* 1966). However, in Cebu and Negros islands, the Paleogene diorite masses associated with mineral deposits have northeast trends. (Atlas Mines, 1967; *Kinkel* et al, 1956).

The basement rocks in Luzon are "crystalline schists and quartzites of pre-Jurassic age, Cretaceous-Paleogene volcanic flows, intercalated cherts, marble and extensive meta-sedimentary rocks", according to *Fernandez & Pulanco* (1966). They added that minor serpentinized peridotites appear to occur in places along faults of late Miocene sedimentary rocks. The undifferentiated schists represent metamorphism of basic flows and sandstone.

The southeastern Luzon area bounded by the Philippine fault and Albay Fault could probably exhibit the characteristics of both the frontal and third arc. This is due to the nearness of the area to the boundary effects of the arc to arc nature of the setting. The Albay Fault in this regard, is taken as the trace of the inferred transform fault dividing the west Luzon and eastern Mindanao subduction zones. In the Paracale-Mambulao district (24,24), according to *Frost* (1959) a granodiorite stock has intruded a sequence of basic and ultrabasic rocks. Unconformably overlying the ultrabasics are a series of Miocene sedimentary rocks. These sedimentaries were in turn overlain unconformably by a sequence of volcanic flows of andesitic composition (Larap Volcanics). The writer has seen the presence of pillow lava with volcanic breccia in the Larap Volcanics indicating submarine deposition.

It is interesting to note that the Larap Volcanics include diorite-andesite-dacite assemblage (dioritic complexes) and basaltic-andesitic submarine flows. It will be recalled that in terms of igneous activity, the Paracale-Mambulao district have both the predominant rock types of the frontal arc (ultrabasic volcanic) and the third arc (intermediate intrusive-intermediate volcanics).

Mineral deposits

The mineral deposits are classified on the basis of their island arc setting, either under the frontal arc or third arc. This classification is summarized in Table 1. The geographic distribution of the mineral deposits is shown in Figure 3. The localities of these mineral deposits do not necessarily indicate their magnitude or economic importance, though some considerations were taken in weighing their economic potential. It must be added that there are numerous small copper-gold deposits that were not included in Fig. 3.

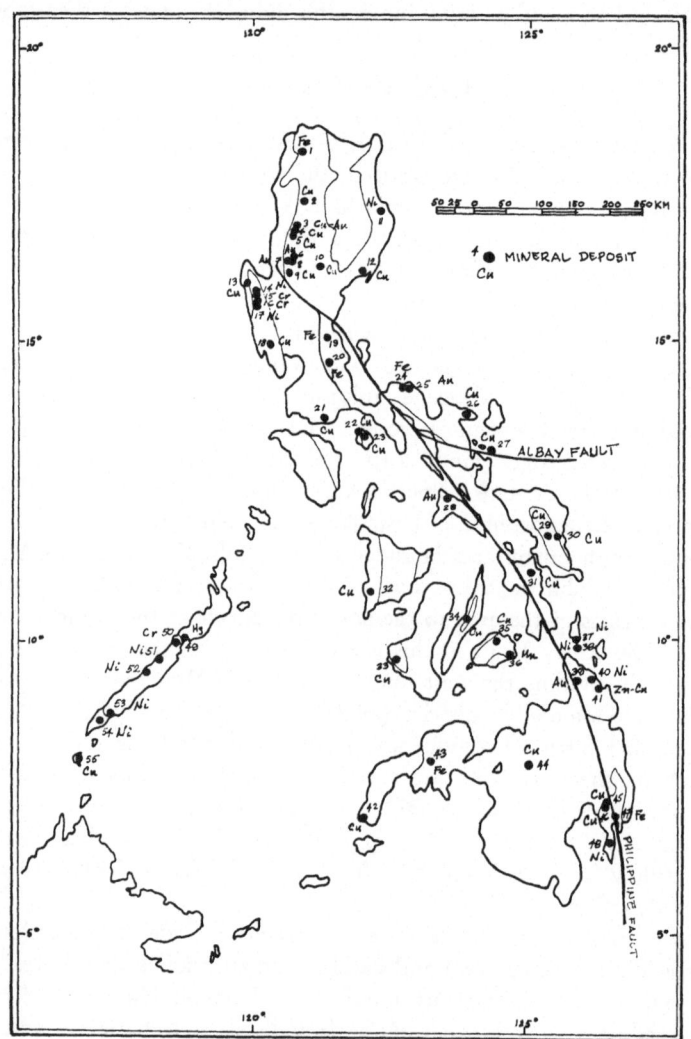

Fg. 3. Geographic Location of Important Mineral Deposits of the Philippines.

Table 1

Classification of mineral deposits of the Philippines

A. Frontal Arc

Map No.	Name & Location of Deposit	Type of Deposit	Geologic Setting	Ore Minerals	Ore Metals	Age
13	Barlo, Pangasinan	Massive sulfide	Inter-basic volcs	Py, cp, sl, bn, te, tn	Cu, Zn, Ag	Paleogene
14	Global Mining, Zambales	Laterite	Ultrabasic	Laterite	Ni, Co	U. Cretaceous (?)
15	Acoje, Zambales	Massive-diss	Ultrabasic-basic	Cr, pn	Cr, Ni, Pt, Pd	U. Cretaceous (?)
16	Coto, Zambales	Massive-diss	Ultrabasic-basic	Cr	Cr	U. Cretaceous (?)
17	Midesco, Zambales	Laterite-diss	Ultrabasic-basic	Lat, pn	Ni	
18	Dizon, Zambales	Diss — stock	Inter-basic	Cp, Py	Cu, Au, Ag	
49	Tagburos, Palawan	Diss	Ultrabasic	Cinnabar	Hg	U. Cretaceous (?)
50	Marsman, Palawan	Massive-diss	Ultrabasic	Cr	Cr	
51	Long Pt., Palawan	Laterite	Ultrabasic	Lat-sap	Ni	
52	Berong, Palawan	Laterite	Ultrabasic	Lat-sap	Ni	
53	Brookes Pt., Palawan	Laterite	Ultrabasic	Gar	Ni	
54	Rio Tuba, Palawan	Laterite	Basic volcs	Lat-gar	Ni	U. Cretaceous Early Eocene
55	Lorraine, Balabac Is	Massive sulfide	ultrabasic	Py, cp, sl, bn	Cu	
32	Bongbongan, Panay	Massive sulfide	Basic volcs ultrabasic	Py, cp	Cu	U. Eocene
26	Sandho, Camarines S.	Massive sulfide	Ultrabasic	Py, cp	Cu	U. Eocene
27	Hixbar, Rapu-Rapu Is	Massive sulfide	metamorphics ultrabasic	Py, cp, sl, chal, cov	Cu, Zn, Au, Ag	
29	Bagacay, Samar	Massive sulfide	Metamorphics ultrabasic	Py, cp, sl, chal, cov	Cu, Zn	
30	Sulat, Samar	Massive sulfide	Inter intrus Inter volcs	Py, cp, bn, sl, gn	Cu, Zn, Pb	M. Miocene
31	Tigbao, Leyte	Massive sulfide	Inter volcs ultrabasic (?)	Py, cp, bn, sl, gn	Cu, Zn	
37	Dinagat Is, Surigao	Laterite-diss	Ultrabasic	Lateritic	Ni, Co, Fe	Miocene (?)
38	Nonoc Is, Surigao	Laterite-diss	Ultrabasic	Lateritic	Ni, Co, Fe	Miocene (?)

97

No.	Locality	Form	Setting	Minerals	Metals	Age
39	Cabadbaran, Agusan	Vein form	Inter volcs (?)		Au, Ag	Miocene (?)
40	Cantillan, Surigao	Laterite	Ultrabasic		Ni	
41	Lanuza, Surigao	Massive sulfide	Inter volcs	Py, cp, sl	Zn, Cu	
45	Sabena, Davao	Diss — stock	Inter intrus		Cu	
46	Masara, Davao	Vein form	Inter volcs &	Mag, sl, cp, py	Cu, Au, Ag	U. Miocene (?)
			Intrus			
47	Mati, Davao	Replacement (?)	Inter volcs &	Mag	Fe	
48	Pujada, Davao	Laterite	Inter intrus	Lat-sap	Ni	Pliocene
			Sed-Inter volcs (?)			
			Ultrabasic			

B. Third Arc

No.	Locality	Form	Setting	Minerals	Metals	Age
1	Lammin, Ilocos Norte	Replacement	Inter volcs	Mag, hem	Fe	Eocene
			Inter intrus			
2	Lubuagan, Kalinga	Diss — stock	Inter intrus	Cp	Cu	U. Miocene
3	Lepanto, Bontoc	Vein form	Inter volcs	En, lu, tn, cp	Au, Cu, Ag	U. Miocene
4	Boneng, Benguet	Diss — stock	Inter intrus	Mag, cp, bn	Cu	U. Miocene (?)
5	Sto. Niño, Benguet	Diss	Inter volcs	Cp, mo, mag	Cu	
			basic intrus			
6	Atok, Benguet	Vein form	Inter volcs	Au, cp, py	Au, Ag, Cu	U. Miocene
7	Thanskgiving, Baguio	Replacement	Inter intrus	Sl, cp, mag	Zn, Au, Ag, Cd	M. Miocene
8	Acupan, Antamok, Baguio	Vein form	Inter volcs &	Au, ag, cp	Au, Ag, Cu	U. Miocene
			intrus			
9	Sto. Tomas II, Benguet	Diss — stock	Inter volcs &	Cp, bn, py, mag, mo	Cu, Au, Ag, Fe	U. Miocene
			intrus			
10	Dupax, Nueva Vizcaya	Massive sulfide	Inter volcs &	Sl, py, cov	Zn, Cu	
			intrus			
11	Palanan, Isabela	Laterite	Ultrabasic	Lat-gar	Ni	
12	Sierre Madre, Quezon	Diss	Inter volcs	Cp, bn, Mn oxide	Cu, Mn	
			Inter intrus			
19	Kamatsihan, Bulacan	Replacement	Inter-basic volcs	Mag	Fe	M. Miocene
20	Sta. Ines, Bulacan	Replacement	Sed-inter volcs	Mag	Fe	U. Miocene
21	Lobo, Batangas	Vein form	Inter volcs	Cp, en, lu, chal, bn	Cu, Au, Ag	
22	CMI, Marinduque	Diss-massive sulfide	Basic volcs	Cp, bn	Cu, Au, Ag	
			Inter intrus			

23	Marcopper, Marinduque	Diss — stock	Basic volcs Inter intrus	Cp, bn	Cu, Au, Ag	M. Eocene
24	Larap, Camarines Norte	Replacement	Metamorphics inter volcs basic-ultra	Mag, cp, mo, gn, py, Uraninite	Fe, Cu, Mo, Sl, Gn, Py, U	
25	Gumaus, Camarines Norte	Vein form	Inter intrus	Au, cp, gn	Au, Ag, Cu, Pb	
28	Aroroy, Masbate	Vein form	Inter intrus Metamorphics	Au	Au	Neogene
33	Sipalay, Negros	Diss — stock	Inter volcs	Cp, bn, mo	Cu, Mo, Au, Ag	Paleocene Early Eocene
34	Lutopan, Cebu	Diss — stock	Inter volcs	Cp, bn, mag, by, mo	Cu, Py, Fe	
35	Talibon, Bohol	Diss	Inter intrus & intrus	Cp, mo	Cu	Paleocene (?)
36	Anda, Bohol	Replacement	Basic volcs-sed	Mn oxide, Pyrolusite	Mn, Zn, Pb	M. Miocene
42	Ayala, Zamboanga City	Vein form	Inter intrus	Py, cp	Cu, Au, Ag	
43	Sibuguey, Zamboanga	Replacement	Sed-inter volcs	Mag	Fe	
44	Malaybalay, Bukidnon	Replacement (?)	Inter intrus Inter volcs	Cp, mag	Cu	

Symbols used

Diss	= disseminated	cp	= chalcopyrite	sap	= saprolite
Stock	= stockwork	sl	= sphalerite	gar	= garnierite
Inter	= intermediate	bn	= bornite	chal	= chalcocite
Volcs	= volcanics	te	= tetrahedrite	mag	= magnetite
Intrus	= intrusive	tn	= tennantite	hem	= hematite
Sed	= sedimentary	pn	= pentlandite	gn	= galena
Ultra	= ultrabasic	cr	= chromite	lu	= luzonite
Py	= pyrite	Lat	= nickel laterite	en	= enargite

Substantial beach and off-shore titaniferous magnetite sands occurrences were not included in the classification of mineral deposits. These deposits are essentially found in western part of Luzon, eastern Leyte and northern Mindanao among others. These magnetite sands were probably derived mainly from the weathering of volcanic rocks *(Santos & Walters, 1971)*. It is believed that these magnetites are erosional products of third arc volcanic rocks. Considering that the magnetite reserve of the western beaches of west Luzon alone is about a few tens of million *(Harrington & Andrews, 1971)*, the total titaniferous magnetite reserve could easily reach more than 200 million metric tons for the entire Philippines.

Frontal Arc Mineral Deposits

The principal type of mineral deposits in the frontal arc in the order of predominance are: (a) lateritic soil (b) massive sulfide (c) massive-disseminated and stockwork (d) minor replacement and vein deposits.

Alpine-type ultrabasic complexes and their associated basic volcanics distinguish the frontal arc rocks. The ultrabasic rocks consist mainly of serpentinized peridotite, gabbro, dunite and minor pyroxenite and dolerite. Tropical weathering in the western parts of Zambales-Palawan and eastern Samar and Mindanao areas has produced thick lateritic blanketing of the ultrabasics. Tropical chemical weathering of the ultramafic rocks has resulted in the economic concentration of nickel-cobalt-iron metals in lateritic soils. The lateritic nickel deposit at Nonoc island (58) alone which is currently being developed have the following mineable reserves (O'Kane, 1971) shown in Table 2.

Table 2

Nickel Reserve of Nonoc island, Philippines

Cut-off Grade % Ni		Area of Orebody (hectares)		Reserve DMT		Grade %						Stripping Ratio
						Ni		Co		Fe		
0.9	:	1 436	:	75 104 000	:	1.22	:	0.10	:	38.4	:	0.61

According to O'Kane the combined nickel mineralization present in the smaller group of islands near Nonoc are greater than the Nonoc reserves. The probable nickel ore reserves of Hinatuan, Awasan, Hanigad and Southern Dinagat (37) island are shown in Table 3.

Table 3

Nickel Reserves of Hinatuan, Awasan, Hanigad and Southern Dinagat Islands
(O'*Kane*, 1971)

Location	Area (Hectares)	Selected cut-off Grade % Ni	Probable Reserves DMT	Grade - %		
				Ni	Co	Fe
Hinatuan Island	529	0.9	27 467 000	1.27	0.14	39.7
Awasan Island	105	1.0	3 922 000	1.23	0.13	40.8
Hanigad Island	50	1.0	3 810 000	1.23	0.09	32.7
Southern Dinagat	1 152	0.9 & 1.0	49 518 000	1.21	0.13	40.2
TOTAL	1 836	—	84 717 000	1.23	0.13	39.8

Table 4

Reserves and Assays of Philippine Nickel Deposits
(Board of Investment Data, 1972)

Type of Ore	Positive Reserves (metric ton)	Grade %	Geologic Reserves (metric ton)	Grade %
Laterite	155	1.24—2.0	418	0.8 —2.0
Garnierite	32	2.06—2.24	25	1.75—2.24
Nickel sulfide	1	0.7	0.6	0.7
Saprolite	—	—	197	1.37—1.53
TOTAL	188		640.6	

Important deposits found in the serpentinized ultramafic rocks are mainly chromite with minor associated nickel-platinum sulfide metals. The chromites are favorably localized in serpentinized dunite or troctolite *(Thayer*, 1960).

The chromite deposits occur in massive-disseminated form with distinctive layering or banding in serpentinized dunite. In Acoje (15), the chromite is associated with nickel-platinum-bearing sulfides *(Irving*, 1950). The Acoje deposit is high in chromium and low in aluminium while the Coto (16) chromite is high in aluminium and magnesium, and low in chromium, *(Fernandez*, 1960). There are a number of chromite bodies distributed in west Luzon and Palawan, however, it appears, that Acoje, Coto and the much smaller, Marsman chromite (50) are the only ones that formed economic mineral deposits.

The massive sulfides are equivalent to the massive and range mineralogically from nearly pure pyrite to intimately mixed pyrite, (chalcopyrite), sphalerite, galena and barite".

Massive sulfide deposits of the cupriferous or "Kuroko" types are often associated with the basic volcanics. The volcanics are mainly spilitic-keratophyric rocks with pillow lavas and associated chert *(John*, 1963; *Bryner*, 1969).

To a lesser degree disseminated-stockwork occur in basic intrusives (?). An example of this is the Dizon copper deposits (18). The author strongly suspects that similar types of deposits may be located along the periphery of major ultramafic complexes, on the side away from oceanic deeps.

Third Arc Mineral Deposits

The third arc type of deposits, in their order of importance, are: (a) disseminated-stockwork (b) vein (c) replacement (d) minor massive sulfide and lateritic soil.

The disseminated-stockwork deposits are mainly of the porphyry copper types which occur in diorite complexes (intermediate intrusive-volcanics) (4, 5, 9, 23, 33, 34). The principal copper mineral is chalcopyrite with associated lesser bornite, pyrite, magnetite, molybdenite, gold and silver.

Quartz veins with gold and silver deposits are found in Northern Luzon (3, 6, 7, 8), southeastern Luzon (28) and Masbate island (28). These gold deposits include the polymetallic type exemplified by Lepanto (3) and Thanksgiving (7). The important copper minerals in Lepanto are enargite, luzonite and tellurides of gold and silver. At Thanksgiving, the minerals present are chalcopyrite, sphalerite and galena with associated gold tellurides. (*Bryner*, 1969).

The gold vein ore deposits at Acupan (8) and Atok (6) are characterized by quartz-calcite-gold mineralization. Generally, gold particles, "0,5 mm to micron sizes, occur in lusters", though gold are also present as tellurides. (BCI, 1967). Base metal sulfides with associated gold occur in minor amounts.

Paracale-Mambulao (southeastern Luzon) and Masbate areas are pre-war gold producing districts. The quartz-gold-bearing sulfides are associated with intermediate intrusive (granodiorite-diorite stock).

Replacement deposits are nearly restricted to iron-base metal sulfide mineralizations (1, 17, 20, 24, 43). *Bryner* (1969) prefers to call this type of deposits as pyrometasomatic deposits while Harrington and *Andrews* (1971) are inclined to call it contact-metasomatic deposits. Host rocks are limestone or calcareous sedimentary rocks adjacent to intermediate volcanics- intermediate instrusives. The chief iron mineral is magnetite and subordinate hematite. In Larap, the massive magnetite orebody is adjacent to a chalcopyrite-molybdenite-magnetite deposit containing uraninite. In Sta. Ines (20), there are also two massive deposits, magnetite and pyrite-pyrrhotite-chalcopyrite. (*Harrington & Andrews*, 1971).

The lateritic soil deposits are primarily aluminous. They are confined mostly in the northern coast of Mindanao. The deposit contains alumina and iron oxides together with other oxides. The average chemical analyses of this aluminous lateritic deposit is: 34.40 % Al_2O_3 9.59 % Fe and 35.23 % SiO_2 (Philippine Bu. of Mines, 1962). Since the aluminous laterite is still non-economic as a source of aluminium raw material, due to high silica content, only the nickel laterite (11) was plotted in Fig. 3.

Period of Mineralization

The age of chromite and massive sulfide deposits could probably span from Upper Cretaceous to Early Eocene. It must be noted that dating of the host rocks of the mineral deposits are inferred from associated fossil-bearing sedimentary series. The age of the nickelferous lateritic soil deposits is still problematical, though Harrington and *Andrews* (1971) suggest that laterization probably commenced during Miocene time in the Nonoc area.

The period of mineralization in the third arc ranged from Upper Paleocene to Upper Miocene time. According to *Wolfe* (1971) the potassium-argon datings of andesites and diorites (dioritic complexes) ranged from Eocene, Oligocene through Miocene. Most of the datings, however, are concentrated during Upper Miocene-Lower Pliocene time. It appears then that this period indicated extensive intermediate intrusive, intermediate volcanic activities and metallization.

It is evident then that there were, at least two major periods of mineralizations involved in the Philippine Island arc system, namely, during Upper Cretaceous to Lower Eocene in the frontal arc and Upper Miocene-Lower Pliocene time in the third arc. There are reasons *(Wolfe,* 1971) to believe, however, that Eocene time is probably another major period of metallization.

Conclusion

In summary, the Philippine mineral deposits are subdivided on the basis of their association with the structural elements of the island arc system. The Philippine archipelago is an "arc to arc" island system with two transform faults. Nickel-cobalt-iron, massive sulfide and chromite deposits occur in ultrabasic instrusive and basic volcanics of the frontal arc.

Copper-gold-silver-magnetite concentrations in disseminated-stockwork, vein and replacement type deposits are typically formed in the third arc. The regional rocks are mainly intermediate intrusive-volcanics (diorite-andesite complexes) intruding Tertiary volcanic-sedimentary sequences.

On the basis of the juxtaposition of the north-south trending Tertiary sediment formations and the northeast trends of the pre-Tertiary igneous-sedimentary rocks of Palawan-Mindoro areas, it is suggested that the date of collision probably occurred at the close of Permian to early Cretaceous. The quartz-rich sediments of Palawan-Mindoro deposited under miogeosynclinal environments indicate acidic source rocks. It may be deduced on the basis of marine fossil evidence that the date of rifting of these quartz-rich formations was probably late Carboniferous time.

Structural and geological evidences support the thesis that the Sulu Sea basin was formed as a result of extensional rifting of the Palawan frontal arc. This implies that a part of the Zamboanga Peninsula was formerly adjacent to Palawan.

References

Atlas Consolidated Mining and Development Corp. (1963): Geology of Atlas open pit mine. — The Philippine Geologist, *17*, p. 100—116.

Bryner, L. (1969): Ore deposits of the Philippines — an introduction to their geology. — Econ. Geol. *64*, pp. 644—666.

Corby, W. (1951): Geology and Oil Possibilities of the Philippines. — Phil. Bur. Mines Tech. Bull. *21*, p. 363.

Daleon, B. A. (1971): A review of oil exploratory drilling in the Philippines (1896—1971). — *In:* 3rd Symposium on Mineral Resources Development and the 18th Mine Safety Conference, Manila, Phil., Nov. 29—30.

Datuin, R. (1972): Phil. Commission on Volcanology, Personal Communications.

Dewey, J. F. & *Bird*, J. M. (1970): Mountain belts and the new global tectonics. — J. Geophys. Res., *75*, pp. 2625—2647.

Dickinson, W. R. (1971): Plate tectonics in geologic history. — Science, *174*, no. 4005, pp. 107—113.

Engineering and Geological Dept., Benguet Consolidated, Inc. (1963): The Baguio gold mining district, Mountain Province, Luzon, Philippines, with special reference to the gold operations of Benguet Consolidated, Inc. (Unpublished).

Fernandez, N. S. (1960): Notes on the geology and chromite deposits of the Zambales Range. — The Philippine Geologist, *14*, pp. 1—8.

Fernandez, J. C. & *Pulanco*, D. (1966): Preliminary report on the reconnaissance geology of northwestern Luzon, Philippines. — Philippine Bur. Mines Report, Manila.

Frost, J. E. (1959): Notes on the genesis of the ore-bearing structures of the Paracale district, Camarines Norte, Philippines. — The Philippine Geologist, *13*, pp. 31—43.

Gervasio, F. C. (1964): A study of the tectonics of the Philippine archipelago. — *In:* 22nd Session of the International Geological Congress in New Delhi, India.

— (1966): The age and nature of orogenesis of the Philippines. — The Philippine Geologist, *20*, pp. 121—140.

Hayes, D. E. & *Ludwig*, W. J. (1967): The Manila Trench and West Luzon Trough — II gravity and magnetic measurements. — Deep-Sea Res., *14*, pp. 545—560.

Harrington, J. F. & *Andrews*, L. E. (1971): Iron ore resources of the Philippines. — Mineral Engineering Magazine, *22*, pp. 14—31.

Irving, E. M. (1949): Review of Philippine basement geology and its problem. — Philippine Jour. Sci., pp. 267—307.

— (1951): Submarine morphology of the Philippine archipelago and its geological significance. — Philippine Jour. Sci., *80*, pp. 55—88

— (1950): A potential nickel-platinum mine at the Acoje chromite mine. — The Philippine Geologist, *5*, pp. 11—22.

Isacks, B., *Oliver*, J. & *Sykes*, L. (1968): Seismology and the new global tectonics. — J. Geophys. Res., *73*, pp. 5855—5899.

John, T. U. (1963): Geology and mineral deposits of east-central Balabac Island, Palawan Province Philippines. — Econ. Geol., *58*, pp. 107—130.

Karig, D. E. (1970): Ridges and basins of the Tonga-Kermadec island arc system. — J. Geophys. Res., *75*, pp. 239—254.

Kintanar, E. R. (1967): Petroleum geology of the Philippines. — Philippine Mining Jour., *19*, pp. 26 (abstract).

Ludwig, W. J., *Hayes*, D. E. & *Ewing*, J. (1967): Bathymetry and sediment distribution, the Manila Trench and West Luzon Trough-I. — Deep-Sea Res., *14*, pp. 533—544.

Melendres, M. N., Jr. (1971): Review of oil exploration in the Philippines. — In 3rd Symposium on Mineral Resources Development and 18th Mine Safety Conference, Manila.

Morante, E. M. (1970): Notes on the seismicity of the earthquakes of 1 August 1968 and 7 April 1970. — Jour. Geological Soc. of the Philippines, *24*, pp. 1—19.

Motegi, M. (1971): On the Palawan trend. — Mineral Engineering Magazine, *23*, pp 30—35.

O'Kane, P. T. (1971): Development of the Surigao nickel project. — In 3rd Symposium on Mineral Resources Development and 18th Mine Safety Conference, Manila.

Philippine Board of Investment (BOI) (1972): (Press release).

Philippine Bur. of Mines (1962): Aluminous laterite deposits in northern Mindanao, potential sources of refractory raw materials. — Philippine Mining Jour., *4*, pp. 20—22.

Philippine Bur. Mines (1963): Geological map of the Philippines, Manila.

Ranneft, T. M. (1960): Reconnaissance geology and oil possibilities of Mindanao. — Bull. Amer. Assoc. Petrol. Geol., *44*, pp. 529—569.

Santos, G. & *Wainerdi*, R. (1969): Notes on the 1965 Taal volcanic eruptions. — Bull. Volcanologique, *33*, pp. 503—529.

Santos, G. & *Walters*, L. J., Jr. (1970): Activation analysis of Philippine magnetite sands. — Presented at the NATO Advance Study Institute on Activation Analysis in Geochemistry and Cosmochemistry, Kjeller, Norway.

Thayer, T. P. (1960): Some critical differences between alpinetype and stratiform peridotite-gabbro complexes. — The Philippine Geologist, *14*, pp. 87—103 (presented at the 21st Int'l Geol. Congr., Copenhagen, Denmark, 1960).

Tupaz, M. H. (1960): A preliminary study of the geology of Philippine copper deposits. — Natural and Applied Sci. Bull., *17*, pp. 283—294.

Wolfe, J. E. (1971): Interpretation of Philippine potassium argon in the Philippines. — In: 3rd National Geological Convention and 1st Field Symposium on Ore Deposits of Philippines, Baguio.

Beziehungen zwischen den metallogenetischen, petrographischen und geochemischen Provinzen der Balkanhalbinsel und Kleinasiens

Stevan *Karamata*[*]

Zusammenfassung

Auf der Balkanhalbinsel und in Kleinasien bestehen je zwei metallogenetische, petrologische und geochemische Provinzen:

a) die Pb-Zn (Sb-Ag) führende metallogenetische Provinz, die genetisch an den sialischen, K-betonten, tertiären Magmatismus, der an Pb angereichert und an Cu arm war, gebunden ist, und der die geochemischen Pb-Provinzen der Dinariden Helleniden, der serbisch-mazedonischen Masse, der östlichen und südlichen Rodopen und der Anatoliden entsprechen, und

b) die Cu-führende metallogenetische Provinz, die genetisch an den kretazisch-tertiären, kalkalkalischen bis K-betonten, hybriden, an Cu angereicherten und Pb armen Magmatismus gebunden ist, wobei die beiden zusammen der geochemischen Cu-Provinz der Balkaniden (Timok-Eruptivkomplex und Srednjegorje) und der Pontiden entsprechen.

Abstract

"Correlation of metallogenetic, petrologic and geochemical provinces on the Balkan Peninsula and in Asia Minor"

On the Balkan Peninsula and in Asia Minor (Fig. 1) in Balkanides (Eastern Serbia and Srednjegorje in Bulgaria) and in Pontides (Turkey) exists a Cu-metallogenetic province associated with andesitic and monzonitic igneous activity of Cretaceous-Tertiary age, i. e. with a Cretaceous-Tertiary hybrid, Ca-alkaline to potassic petrographic province (Fig. 3). More to the west and south the Pb-Zn-Sb-Ag metallogenetic province of Dinarides, Hellenides, the Serbo-Mazedonian

[*] Prof. Dr.-Ing. S. *Karamata*, Rudarsko-geološki fakultet, Djusina 7, 11 000 Beograd, Yugoslavia.

Mass, Eastern and Southern Rodops and Anatolides occurs associated with a K-enriched granodioritic (intrusive to extrusive) igneous activity of Tertiary age, i. e. with a Tertiary, potassic, sialic petrographic province (Fig. 2).

Studies of Pb and Cu contents in igneous rocks of both provinces were performed (about 250 samples were analysed for Pb, Cu and K, about 50 published data for Pb and Cu contents and another 100 published data for only Pb or Cu were used). The analysed rock samples were not collected systematically, so they do not represent the abundance of different kinds of rocks, but anyway they give a regional picture.

In comparison with the mean Pb, resp. Cu contents of analogous igneous rocks an enrichment in Pb and Cu is evident, about 80 % of investigated samples show higher Pb, resp. Cu contents than the average contents of these elements in analogous rocks (Fig. 4). This is probably reflected in the high intensity and extensivity of mineralisation in the Balkan Peninsula and in Asia Minor.

The igneous rocks of Dinarides, the Serbo-Mazedonian Mass and Anatolides are enriched in Pb, but poorer in Cu repsectively to the igneous rocks of Balkanides and Pontides. On the other hand the igneous rocks of Balkanides and Pontides are poor in Pb but rich in Cu. In the same metallogenetic and petrologic province going eastwards the Cu and the Pb contents increase, but the rise of Cu contents is faster. Thus the Tertiary igneous rocks of Dinarides, Hellenides(?), the Serbo-Mazedonian Mass, Southern and Eastern Rodops and Anatolides belong to a Pb-geochemical province, and the Cretaceous-Tertiary igneous rocks of Balkanides and Pontides to a Cu-geochemical Province. These two geochemical provinces coincide perfectly with the metallogenetic Lead- and Copper-Provinces established by *W. E. Petrascheck* 20 years ago.

The difference in metallogenetic characteristics, as in petrologic and geochemical properties of igneous rocks from both areas are probably the result of different origin of magma: in the Pb metallogenetic and geochemical province they were produced by melting of a continental crust primarily enriched in Pb and poor in Cu (the Palaeozoic and Jurassic granitic rocks in this area enriched in Pb too). In the Cu metallogenetic and geochemical province the magmas originated by melting of subcrustal material (in roots of a rift system or in a subduction zone) and their mixing with material of the continental crust.

The concentration of Pb contents and Pb/Cu ratios of Tertiary igneous rocks, as well as of Paleozoic and Jurassic granitic rocks in the left parts of the Pb/Cu diagrams (Fig. 4 and 7) indicate that the continental crust which gave these magmas was primary enriched in Pb and very poor in Cu. The line "Pb ppm $= \frac{3}{2}$ Cu ppm — 5 ppm" represents therefore the boundary of the primary Pb contents and of primary Pb/Cu ratios in the continental crust of this area. The increase of Cu, followed by a slighter rise of Pb in the Pb province towards east (in Asia Minor) is probably provoked by contamination of magmas by subcrustal material (strong fracturing in Anatolides and young basaltic activity, as well as the absence of a median mass between Anatolides and Pontides).

Der Bergbau auf der Balkanhalbinsel und in Kleinasien war schon seit der Antike sehr aktiv und ist es bis in unsere Zeit geblieben, deshalb sind lagerstättenkundliche Angaben schon reichlich vorhanden, wodurch man sich über die metallogenetische Provinzen dieser Gebiete ein gutes Bild machen kann. Über die Beziehungen dieser Lagerstättenbildungen zu irgendeinem Magmatismus sind die Angaben viel seltener, und über geochemische Eigenschaften dieser Gebiete, bzw. der Gesteine, die genetisch mit der Lagerstättenbildung verbunden sind, waren sie bis vor 5 Jahren kaum zu finden.

Eine Synthese der jetzt bestehenden Angaben über metallogenetische, petrographische und geochemische Eigenschaften der Balkanhalbinsel und Kleinasiens ist in der Serbischen Akademie der Wissenschaften und Künste im Druck (*Karamata*, 1973 — im Druck), hier wird nur eine Zusammenfassung dieser Arbeit, mit einigen Betrachtungen über Beziehungen zwischen einzelnen Provinzen angegeben.

Die metallogenetischen Untersuchungen, die zu einer regionalen Einteilung führten, waren in Jugoslawien von A. *Cissarz* (1956) und S. *Janković* (1967), in Bulgarien von J. *Jovčev* (1965), in der Türkei von A. *Gümüs* (1970) und für den ganzen balkanisch-anatolischen Bereich schon 1955 von W. E. *Petrascheck* durchgeführt worden. Nach allen diesen Angaben kann man zwei junge, kretazisch-tertiäre metallogenetische Provinzen unterscheiden: eine Pb-Zn (-Sb-Ag) Provinz in den Dinariden, Helleniden, der serbisch-mazedonischen Masse und Rodopen, wie auch weiter in den Anatoliden, und eine Cu-führende in den Balkaniden (Ost-Serbien und Srednjegorje in Bulgarien) und in den Pontiden. Diese metallogenetischen Provinzen, sowie deren Beziehungen zu den geotektonischen Einheiten sind in Abb. 1 dargestellt.*

In der metallogenetischen Cu-Provinz befinden sich z. B. die Lagerstätten Majdanpek, Bor u. a. in Jugoslawien, der Panadžurište-District mit Medet und die Lagerstätte Rosen bei Burgas in Bulgarien, dann Lahanos und Murgul in Türkei, und in der weiteren Fortsetzung das Zagros Gebiet mit Sar Cheshmeh in Iran, sowie eine große Anzahl kleinerer Cu-Vorkommen. Die Pb-Zn Vererzungen sind dagegen selten und ohne Bedeutung.

In der metallogenetischen Pb-Zn (Sb-Ag) Provinz befinden sich z. B. die Lagerstätten Srebrenica, Rudnik, das Kopaonik-Gebiet mit Trepča, dann Lece, Sase und Zletovo in Jugoslawien, Laurion in Griechenland, das Osogovo und Madan-Gebiet in Bulgarien, weiter Balya Maden, Kaleköy und Bolkadag in Türkei, und eine große Zahl anderer größerer oder kleinerer Vorkommen von

---

* Es soll bemerkt werden, daß die Grenze der Anatoliden und Pontiden, bzw. der Pb-Zn- und der Cu-Provinz im westlichen Kleinasien von den von A. *Gümüs* (1970) gemachten Angaben abweicht. Sie wurde wegen besserer Korrelation mit den westlicher gelegenen Gebieten nördlicher gestellt, und entspricht der Südgrenze der Verbreitung der kretazischen vulkanogen-sedimentären Fazies (mit den Cu-Vererzungen), die als charakteristisch für die Balkaniden-Pontiden und für die metallogenetische Cu-Provinz angenommen wird. Die in den meisten der letzten Arbeiten angegebene Grenze, die südlich vom Marmara-Meer liegt, entspricht nach unserer Auffassung jüngeren geotektonischen Vorgängen, die weder für die Cu-Metallogenese, noch für die jüngeren Pb-Zn-Vererzungen eine genetische Bedeutung hatten.

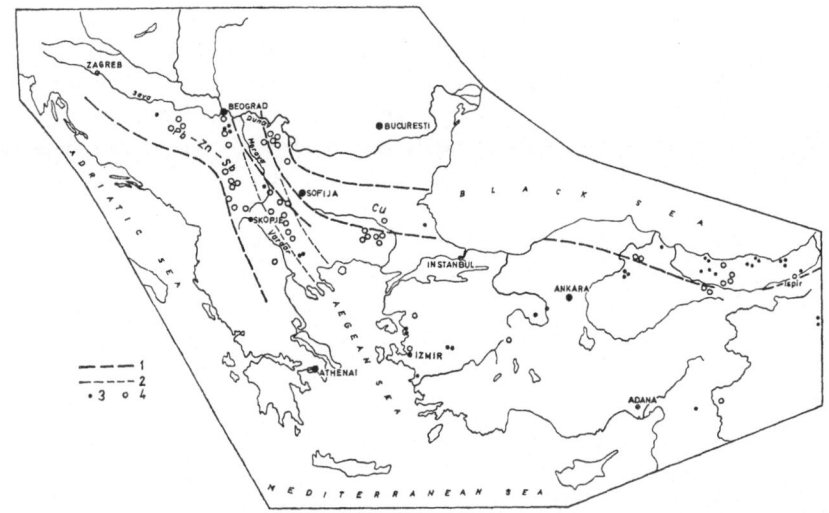

Abb. 1. Metallogenetische, petrographische und geochemische Provinzen kretazischen und tertiären Alters auf der Balkanhalbinsel und in Kleinasien.
Zeichenerklärung: 1 — Grenzen der metallogenetischen Provinzen; 2 — Grenzen der geotektonischen Einheiten; 3 — geochemisch untersuchte Einzelproben; 4 — mehr als 3 geochemisch untersuchte Proben.

Fig. 1. Cretaceous-Tertiary metallogenetic and petrographic provinces and geotectonic units on the Balkan peninsula and in the Asia Minor.
1 — boundary of metallogenetic province; 2 — boundary of geotectonic unit; 3 — investigated sample; 4 — group of more than 3 geochemically investigated samples.

Pb-Zn Erzen. In derselben Provinz sind die Antimonitvererzungen von Zajača, Lisa, Bujanovac, Lojane und Alšar (Jugoslawien), sowie die von Ivrindi, Yeni Gümüs und Turhal (in Türkei) genetisch mit der Pb-Zn Mineralisation verbunden, wenn wir nur die wichtigsten erwähnen. Junge (kretazisch-tertiäre) Kupfer-Vererzungen sind in diesem Gebiet äußerst selten und ohne Bedeutung.

Auf der Balkanhalbinsel und in Kleinasien sind also zwei junge metallogenetische Provinzen klar zu unterscheiden: eine kretazisch-tertiäre Kupfer-führende Provinz der Balkaniden und der Pontiden, und eine Pb-Zn (Sb-Ag)-führende tertiäre metallogenetische Provinz der Dinariden, Helleniden, der serbisch-maze-donischen Masse, der Rodopen und der Anatoliden.

Zusammenfassende Überblicke der kretazisch-teriären magmatischen Tätigkeit aus diesem Gebiet wurden bisher von S. *Karamata* (1962) für die Gesteine der Dinariden und der serbisch-mazedonischen Masse, von R. *Ivanov* (1966) für die Gesteine der Rodopen, und von *Karamata* & al. (1967) für die Gesteine der Balkaniden Jugoslawiens gefaßt.

Die Pb-Zn (Sb-Ag)-Mineralisation in den oben erwähnten Gebieten ist genetisch mit einem intrusiven bis extrusiven Magmatismus tertiären Alters verbunden. Dieser Magmatismus ist granodioritisch-quarzmonzonitisch in Intrusivfazies, und andesitisch-dazitisch-quarzlatitisch in der Extrusivfazies. Nach Osten (in der Türkei) folgt ihm ein starker trachybasaltischer bis basaltischer Vulkanismus. Mit Ausnahme dieses basaltischen Vulkanismus ist der Charakter dieser magmatischen Tätigkeit kalibetont (Abb. 2) und zu den späteren Produkten hin immer K-reicher. So ein Magmatismus kann seinem Charakter nach der Aufschmelzung der tiefen Teile der kontinentalen (sialischen) Kruste entstammen.

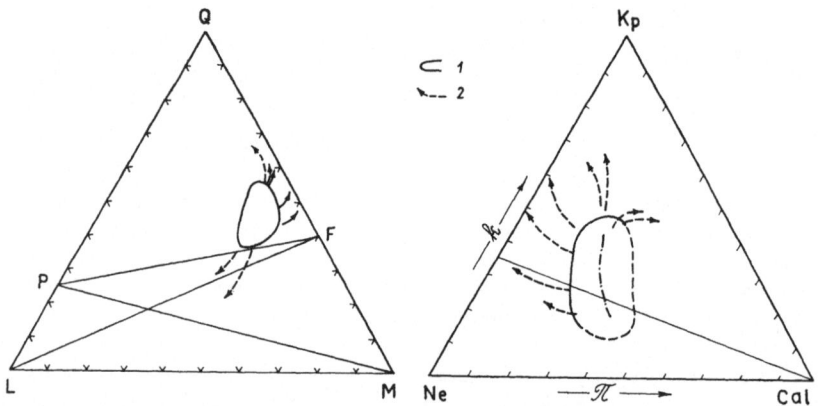

Abb. 2. Petrochemischer Charakter der tertiären Eruptivgesteine der Dinarden und der serbisch-mazedonischen Masse dargestellt in den QLM- und kπ-Diagrammen nach Niggli. 1 — Bereich der maximalen Konzentration der Daten; 2 — Differentiationstendenzen.

Fig. 2. QLM- and kπ-diagrams (after Niggli) illustrating the petrochemical characteristics of Tertiary igneous rocks of Dinarides and of the Serbo-Mazedonian massif. 1 — main field; 2 — differentiation tendency.

Die Cu-Metallogenese ist genetisch gebunden an einen andesitischen (I. Phase) bis andesitisch-basaltischen (II. Phase) Vulkanismus, der sich später in einen trachy-andesitisch-latitischen (III. Phase) bzw. granodioritisch-monzonitischen Magmatismus entwickelt. Der chemische Charakter dieses Magmatismus ist in der Abb. 3 dargestellt. Dieser Magmatismus entspricht den Schmelzen, die durch Aufschmelzung von subkrustalem Material und deren Vermischung mit Krustenmaterial entstanden sind. Diese magmatische Tätigkeit begann in der Oberen Kreide und setzte sich bis ins Paläogen (in den Pontiden dauert sie auch länger) fort.

Es ist also vollkommen berechtigt von zwei jungen petrologischen Provinzen zu sprechen: einer K-betonten, sialischen, tertiären magmatischen Provinz der Dinariden, Helleniden, der serbisch-mazedonischen Masse, der Rodopen und der Anatoliden, die genetisch mit der Pb-Zn (-Sb-Ag)-Mineralisation dieser Gebiete verbunden ist, und demgegenüber einer kalkalkalischen bis, später, K-betonten, hybriden, kretazisch-teritären petrologischen Provinz der Balkaniden (Timok-Eruptivkomplex und Srednjegorje) und der Pontiden, die genetisch mit der Cu-Mineralisation verbunden ist.

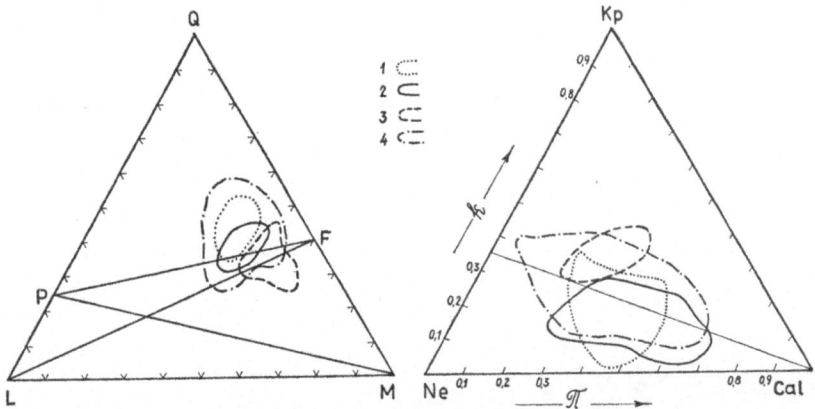

Abb. 3. Petrochemischer Charakter der kretazisch-tertiären Eruptivgesteine des Timok-Eruptivkomplexes in Ostserbien dargestellt in den QLM- und kπ-Diagrammen nach Niggli (nach Karamata u. a., 1967).
1 — Feld der Ergußgesteine der I. Phase; 2 — Feld der Ergußgesteine der II. Phase; 3 — Feld der Ergußgesteine der III. Phase; 4 — Feld der Intrusivgesteine.

Fig. 3. QLM- and kπ-diagrams (after Niggli) illustrating the petrochemical characteristics of Cretaceous-Tertiary igneous rocks of the Timok eruptive area (from Karamata a. coauthors, 1967).
1 — volcanics of the I phase; 2 — volcanics of the II phase; 3 — volcanics of the III phase; 4 — intrusive rocks.

Ein besonderes Verhalten haben die jungen Effusivgesteine der westlichen Teile der Balkaniden in Jugoslawien (die Ridan-Krepoljin Zone) und der östlichen Rodopen in Bulgarien gezeigt: die ersteren sind den Gesteinen der Dinariden ähnlich, während die zweiten den Gesteinen des Srednjegorje nahe kommen.

Geochemische Untersuchungen der Eruptivgesteine der Balkanhalbinsel und Kleinasiens begannen erst vor einigen Jahren, etwa zur selben Zeit in Bulgarien und Jugoslawien und erst vor 2—3 Jahren in der Türkei. Diese Arbeiten waren jedoch auf engere Gebiete, oder Gesteinsgruppen bezogen. Die erste regionale Synthese wurde für die Dinariden 1967 (Cuturic & Karamata) ausgearbeitet, und eine umfassende ist im Druck (Karamata).

Da Kupfer und Blei charakteristische Elemente für die Metallogenese in diesem Gebiet sind, wurde besondere Aufmerksamkeit auf dieselben gerichtet. Heutzutage verfügen wir mit ca. 250 Gesteinsanalysen auf Pb, Cu und K, mit ca. 50 Analysen der Gesteine auf Pb und Cu, und mit ca. 100 auf nur Pb oder nur Cu über Bestimmungen in Gesteinen, die eine Zusammenfassung schon erlauben.

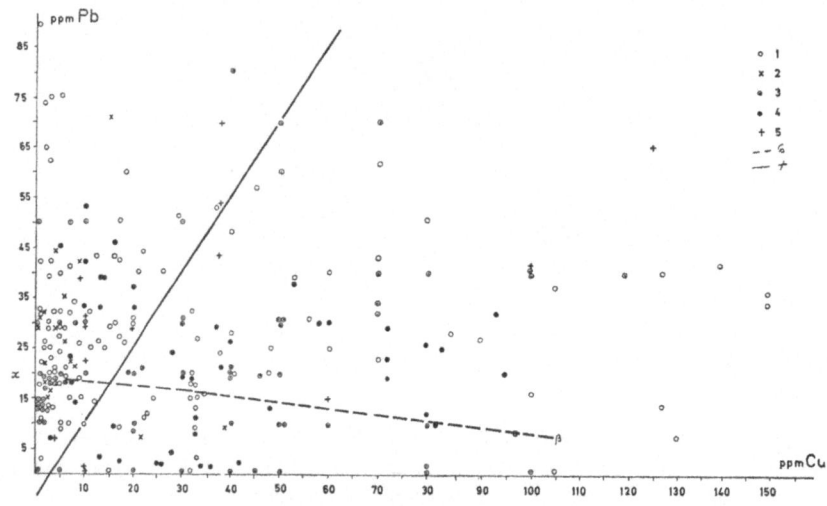

Abb. 4. Blei- und Kupfergehalte in oberkretazischen und tertiären Eruptivgesteinen Jugoslawiens, Bulgariens und der Türkei.

1 — Ergußgesteine der Dinariden und Anatoliden; 2 — Intrusivgesteine der Dinariden und Anatoliden; 3 — Ergußgesteine der Ridan-Krepoljin-Zone (Ostserbien), der östlichen Rodopen und der Übergangszone zwischen den Anatoliden und Pontiden; 4 — Ergußgesteine des Timok-Eruptivkomplexes, des Srednjegorje in Bulgarien und der Pontiden; 5 — Intrusivgesteine des Timok-Eruptivkomplexes, des Srednjegorje und der Pontiden; 6 — Linie der mittleren Pb/Cu-Verhältnisse in Eruptivgesteinen (für Granite nach *Turekian* & *Wedepohl* 1961, für Diorite und Basalte nach *Vinogradov* 1962); 7 — die „Grenze der primären Pb-Gehalte und der primären Pb/Cu-Verhältnisse".

Fig. 4. Pb/Cu diagram for the Upper Cretaceous and Tertiary igneous rocks of Yugoslavia, Bulgaria and Turkey.

1 — volcanics of Dinarides and Anatolides; 2 — intrusives of Dinarides and Anatolides; 3 — volcanics of the Ridan-Krepoljin zone in Eastern Serbia, of Eastern Rodops in Bulgaria and of the bordering zone of Anatolides and Pontides in Turkey; 4 — volcanics of the Timok eruptive area, of Srednjegorje in Bulgaria and of Pontides; 5 — intrusives of the Timok eruptive area, of Srednjegorje and of Pontides; 6 — the Pb/Cu ratio in igneous rocks on the basis of average Pb and Cu contents in granites (after *Turekian* & *Wedepohl* 1961), diorites and basalts (after *Vinogradov* 1962); 7 — the "boundary line of primary Pb content and primary Pb/Cu ratio".

112

In der Abb. 4 sind alle bestimmte Pb- und Cu-Gehalte der Gesteine auf einem Diagramm angegeben um diese Gehalte mit den mittleren Gehalten derselben Elemente in Eruptivgesteinen der Erdkruste (nach *Turekian* & *Wedepohl*, 1961, und nach *Vinogradov*, 1962) zu vergleichen. Es ist leicht ersichtlich, daß mehr als 80 % aller Proben der Dinariden, der serbisch-mazedonischen Masse und der Anatoliden im Vergleich mit den mittleren Gehalten des Pb und Cu in Eruptivgesteinen der Erdkruste erhöhte Pb-Gehalte und meistens niedrigere Cu-Gehalte aufweisen. Die Gesteine des Timok-Eruptivkomplexes und der Pontiden sind dagegen etwas ärmer an Blei aber reicher an Kupfer.

Die Gesteine des untersuchten Gebietes sind also in einer Zone an Blei, in einer anderen an Kupfer angereichert, womit man auch die reiche Erzführung wahrscheinlich erklären kann.

Das zweite wichtige Ergebnis ist die starke Konzentrierung der Angaben, besonders für die Gesteine der Dinariden, im linken Teil des Diagramms. Dieses Gebiet ist mit der Linie

$$Pb = \frac{3}{2} \; Cu - 5 \; \text{(in ppm)}$$

begrenzt. Diese Linie haben wir „die Grenze der primären Pb-Gehalte und der primären Pb/Cu-Verhältnisse" genannt und ihre Bedeutung werden wir später berücksichtigen.

Die Pb-, Cu- und K-Gehalte der Eruptivgesteine wurden in die K/Pb-, K/Cu- und Pb/Cu-Diagramme eingetragen und danach wurden in den Diagrammen die Gebiete mit maximaler Konzentration der Daten für einzelne geotektonische Einheiten ausgeschieden (Abb. 5, 6 und 7). Aus diesen Diagrammen kann man über geochemische Eigenschaften der kretazisch-tertiären Gesteine dieses Gebietes folgendes schließen:

— die Gesteine der Dinariden sind in Bezug zu den mittleren Pb- und Cu-Gehalten der Eruptivgesteine (nach *Turekian* & *Wedepohl* 1961 und nach *Vinogradov* 1962) an Blei angereichert, an Kupfer aber sehr arm.

— Die Gesteine der Dinariden sind an Pb reicher als die Gesteine des Timok-Eruptivkomplexes, und die Gesteine der Anatoliden sind Pb-reicher als die Gesteine der Pontiden. Dagegen sind die Gesteine des Timok-Eruptivkomplexes und der Pontiden reicher an Cu als die Gesteine der Dinariden bzw. der Anatoliden.

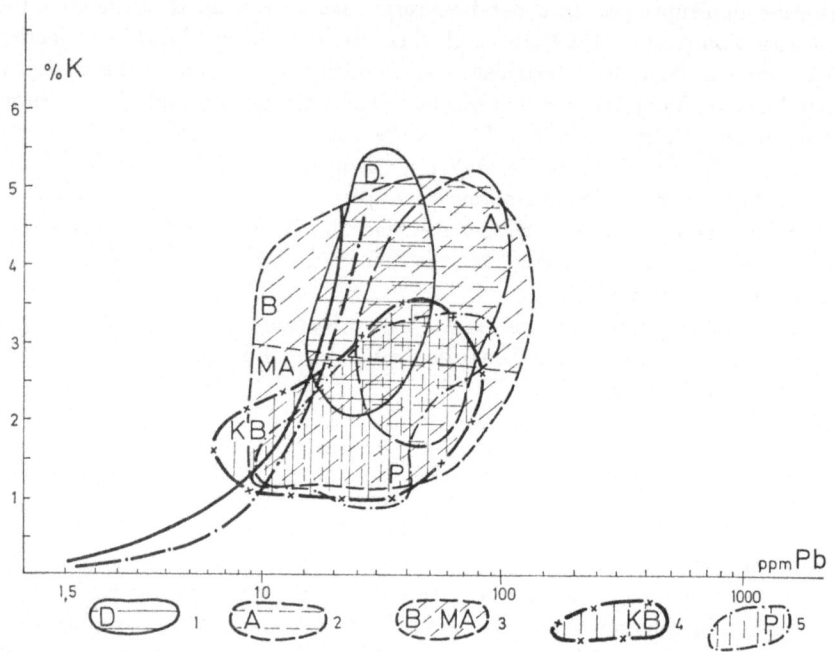

Abb. 5. Bereiche der maximalen Konzentration der Angaben im K/Pb-Diagramm für die oberkretazischen und tertiären Eruptivgesteine der Dinariden (1), der Anatoliden (2), Bulgariens (3, B — der Borovica Serie, MA — der Momčilgrad-Ardino Serie vulkanischer Gesteine), des Timok-Eruptivkomplexes, bzw. der Karpatho-Balkaniden Ostserbiens (4) und der Pontiden (5). Die volle dicke Linie gibt das K/Pb-Verhältnis nach den Angaben der mittleren Pb- und K-Gehalte in den Hauptvertretern der Eruptivgesteine von *Turekian* & *Wedepohl* (1961), die dicke strichpunktierte Linie zeigt dasselbe Verhältnis nach den Angaben von *Vinogradov* (1962).

Fig. 5. Areas with maximal concentration of data in K/Pb diagram for Upper Cretaceous and Tertiary igneous rocks of Dinarides (1), Anatolides (2), Bulgaria (3, B — Borovica volcanic series, MA — Momčilgrad-Ardino volcanic series), Timok eruptive area, i. e. Carpatho-Balkanides of Eastern Serbia (4) and of Pontides (5). Full line defines the K/Pb ratio based on average K and Pb contents in main types of igneous rocks after *Turekian* & *Wedepohl* (1961), dashed line with dots define the same ratio after data given by *Vinogradov* (1962).

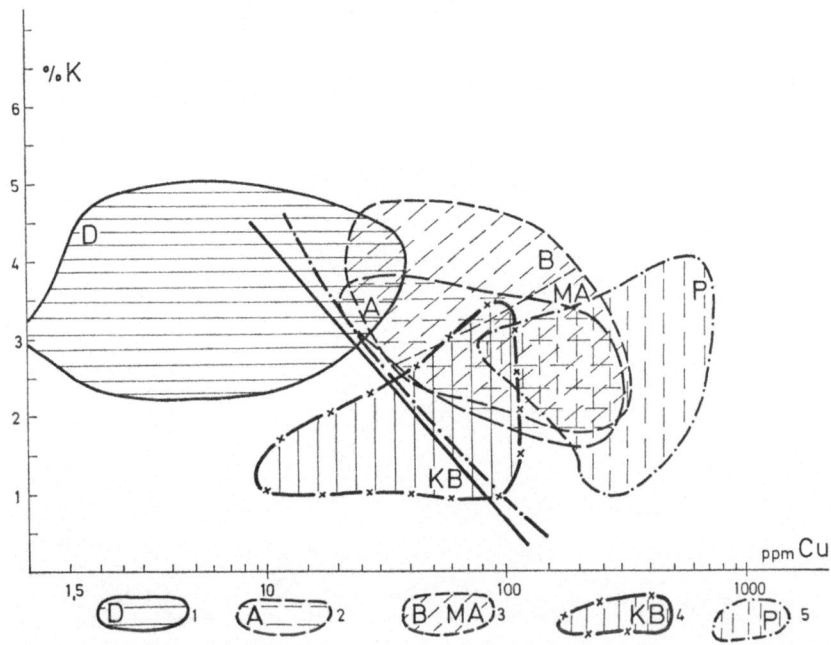

Abb. 6. Bereiche der maximalen Konzentration der Angaben im K/Cu-Diagramm für die oberkretazischen und tertiären Eruptivgesteine der Dinariden (1), der Anatoliden (2), Bulgariens (3, B — der Borovica Serie, MA — der Momčilgrad-Ardino Serie vulkanischer Gesteine), des Timok-Eruptivkomplexes, bzw. der Karpatho-Balkaniden Ostserbiens (4) und der Pontiden (5). Volle und strichpunktierte Linien zusammengestellt für K- und Cu-Gehalte analog wie in Abb. 5.

Fig. 6. Areas with maximal concentration of data in K/Cu diagram for Upper Cretaceous and Tertiary igneous rocks of Dinarides (1), Anatolides (2), Bulgaria (3, B — Borovica volcanic series, MA — Momčilgrad-Ardino volcanic series), Timok eruptive area, i. e. Carpatho-Balkanides of Eastern Serbia (4) and of Pontides (5). Full line define the K/Cu ratio based on average K and Cu contents in main types of igneous rocks after *Turekian* & *Wedepohl* (1961), dashed line with dots define the same ratio after data given by *Vinogradov* (1962).

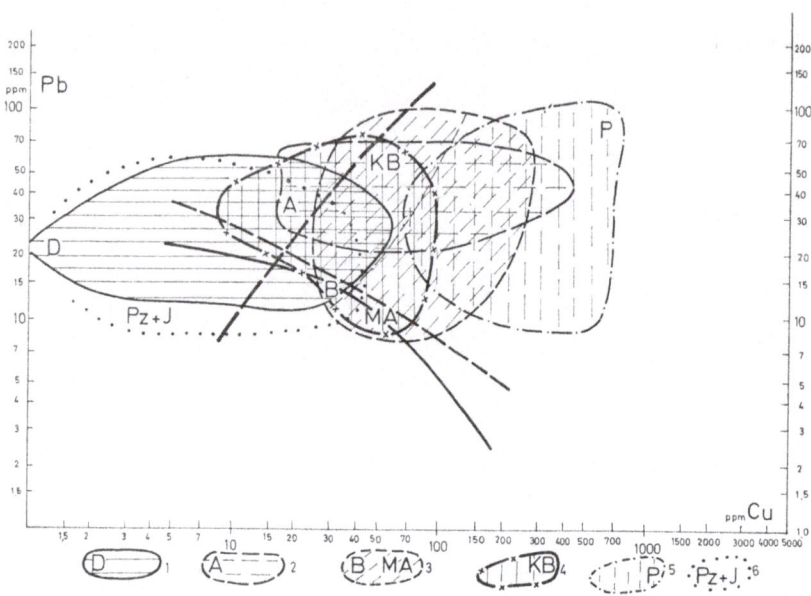

Abb. 7. Bereiche der maximalen Konzentration der Angaben im Pb/Cu-Diagramm für die oberkretazischen und tertiären Eruptivgesteine der Dinariden (1), der Anatoliden (2), Bulgariens (3, B — der Borovica Serie, MA — der Momčilgrad-Ardino Serie vulkanischer Gesteine), des Timok-Eruptivkomplexes, bzw. der Karpatho-Balkaniden Ostserbiens (4) und der Pontiden (5). Eingezeichnet ist auch der Bereich, in dem sich die Angaben über die Pb- und Cu-Gehalte in den paläozoischen und jurassischen Graniten der Dinariden befinden (6). Die volle dicke Linie gibt das Pb/Cu-Verhältnis nach den Angaben der mittleren Pb- und Cu-Gehalte in den Hauptvertretern der Eruptivgesteine von *Turekian* & *Wedepohl* (1961) an, die dicke strichpunktierte Linie zeigt dasselbe Verhältnis nach den Angaben von *Vinogradov* (1962). Die dicke gestrichelte Linie stellt die „Grenze der primären Pb-Gehalte und der primären Pb/Cu-Verhältnisse" dar.

Fig. 7. Areas with maximal concentration of data in Pb/Cu diagram for Upper Cretaceous and Tertiary igneous rocks of Dinarides (1), Anatolides (2), Bulgaria (3, B — Borovica volcanic series, MA — Momčilgrad-Ardino volcanic series), Timok eruptive area, i. e. Carpatho-Balkanides of Eastern Serbia (4) and of Pontides (5). Area of Pb/Cu ratios of Paleozoic and Jurassic granites of Dinarides is indicated (6). Full line defines the Pb/Cu ratio based on average Pb and Cu contents in main types of igneous rocks after *Turekian* & *Wedepohl* (1961), the dashed line with dots defines the same ratio after data given by *Vinogradov* (1962). The dashed, diagonal line is the "boundary line of primary Pb content and of primary Pb/Cu ratio".

116

Man kann also sprechen von einer geochemischen Pb-Provinz der Dinariden, der serbisch-mazedonischen Masse, der Helleniden (keine Angaben) und der Anatoliden, die sich durch erhöhte Blei- und niedrigere Kupfer-Gehalte auszeichnet, und von einer geochemischen Cu-Provinz der Balkaniden (Timok-Eruptivkomplex und Srednjegorje) und der Pontiden, die sich durch Anreicherung an Kupfer und etwas niedrigere Blei-Gehalte kennzeichnet. Nach Osten werden die Gesteine derselben Provinzen langsam an Blei und rasch an Cu angereichert aber die genannten Provinzen kann man leicht eine von der anderen unterscheiden.

Auf der Balkanhalbinsel und in Kleinasien entspricht also die geochemische Pb-Provinz vollkommen der sialischen, K-betonten, tertiären magmatischen Provinz und der metallogenetischen Pb-Zn (Sb-Ag) Provinz, und die geochemische Cu-Provinz der kalkalkalischen bis K-betonten, hybriden, kretazisch-tertiären magmatischen Provinz und der metallogenetischen Cu-Provinz.

Die Gesteine der Dinariden (Abb. 7) befinden sich meistens links von der „Grenze der primären Pb-Gehalte und der primären Pb/Cu-Verhältnisse", die Gesteine des Timok-Eruptivkomplexes und der Anatoliden größtenteils rechts von derselben Linie, und die Gesteine der Pontiden sind weit rechts. Die Gesteine der Rodopen haben eine mittlere Lage. Es ist wichtig, daß die paläozoischen und jurassischen Granite der Dinariden im selben Gebiet des Pb/Cu-Diagramms wie die jungen Eruptivgesteine der Dinariden und der serbisch-mazedonischen Masse liegen.

Die letzten Ergebnisse kann man leicht erklären wenn man annimmt, daß alle diese Gesteine der Dinariden der Aufschmelzung einer selben kontinentalen Kruste entstammen. Wir können dann auch die Pb-Gehalte dieser Gesteine und das Pb/Cu-Verhältnis in ihnen, nachdem sie vom Paläozoikum bis Quartär unverändert blieben, als ein Maß des primären Pb-Gehaltes und des primären Pb/Cu-Verhältnisses annehmen. Die vorher erwähnte Linie stellt die Grenze des Bereiches dieser primären Pb-Gehalte und der primären Pb/Cu-Verhältnisse in der kontinentalen Kruste dieses Gebietes dar. Nach Osten, in den Anatoliden, wo die kontinentale Kruste im Tertiär stark zerstückelt und mit dem Material des Oberen Mantels durchtränkt wurde, kam es zu einer Cu-Anreicherung.

Die Gesteine des Timok-Eruptivkomplexes, die subkrustalen Magmen entsprechen (Rift-System nach *Andric* & al., 1972, oder subkrustale Aufschmelzung in einer „subduction zone"?), waren schon primär Cu-reich, aber gegen Osten, in den Pontiden, wo starke Ozeanisierung noch in jüngster Zeit vorging und wo der Einfluß des Materials des Oberen Mantels besonders stark war, wurde Kupfer noch angereichert.

Es bleibt offen die Frage der Pb-Anreicherung in den mehr östlich gelegenen Gebieten. Es ist möglich, daß diese, wenn auch schwache Pb-Anreicherung, auch mit der Neuzufuhr des Materials aus dem Oberen Mantel verbunden ist.

Blei ist also ein Element der kontinentalen Kruste, das durch geologische Prozesse (Magmenbildung, Erstarrung der Magmen, Remobilisierung usw.) umgelagert wird und lokal konzentriert sein kann. Es ist besonders in seiner Migration an sialische Magmen gebunden und, wahrscheinlich, nur unwesentlich aus dem Oberen Mantel zugeführt worden. Kupfer dagegen ist ein Element des Oberen Mantels oder der ozeanischen Kruste, das bei geotektonischen Vorgängen in kleineren oder stärkeren Massen in die kontinentale Kruste eingeführt sein kann. Kupfer ist also an subkrustale Magmen gebunden und ist in Gebieten mit solchen, schwach bis stärker hybridisierten Magmen angereichert. In die kontinentale Kruste wurde es eingeführt nur wo sie stark zerstückelt und mit dem Material des Oberen Mantels durchtränkt war.

Literaturhinweise

Agiorgitis, G. (1967): Zur Geochemie einiger seltener Elemente in basaltischen Gesteinen. — Tschermaks Min. Petr. Mitteilungen, *B. XII*, H. 2—3, 204—229, Wien.

Aleksić, V., *Pantić*, N. & *Kalenić*, M. (1971): Razmatranje nekih tektonskih procesa u Srbiji u vezi sa tektonikom ploča ili novom globalnom tektonikom. — Glasnik Prirodnjačkog muzeja, A, *26*, 83—102, Beograd.

Andrić, B., *Antonijević*, I., *Grubić*, A., *Dragašević*, T., *Djordjević*, M. & *Terzić*, M. (1972): Analiza gradje timočkog rovsinklinorijuma u svetlosti novih geoloških i geofizičkih podataka. — III savetovanje o bakru RTB, Bor.

Arnaudova, R., *Arnaudov*, B. & *Pavlova*, M. (1971): Razpredelenie na olovoto, barija i stroncija v kalievite feldšpati ot Osogovskite tercierni magmatiti. — Izvestija na geol. institut BAN, Ser. geol. min. i petr., *20*, 5—20, Sofija.

Arsenijević, M. & *Pešić*, D. (1964/65): Rare and dispersed elementes in the Suvi Dol granitoid complex (Stara Planina Mountain). — Vesnik Zavoda za geol. i geof. istra. A, *XXII/XXIII*, 77—115, Beograd.

Cissarz, A. (1956): Lagerstätten und Lagerstättenbildung in Jugoslawien. — Rasprave Zavoda za geol. i geof. istra. NRS, *VI*, Beograd.

Ćuturić, N. & *Karamata*, S. (1967): Die Bleigehalte in tertiären magmatischen Gesteinen Jugaslawiens. — Geologicky sbornik — Geologica Carpathica, *VIII/1*, 27—38, Bratislava.

Ćuturić, N., *Kafol*, N. & *Karamata*, S. (1968): Lead contents in K-feldspars of young igneous rocks of the Dinarides and neighbouring areas. — Origin and distribution of elements, 739—747, Oxford-New York.

Deleon, G. (1958): Trace elements content of granits from Serbia and Macedonia and some geochemical implications. — Vesnik Zavoda za geol. i geof. istraživanja NRS. *XVI*, 167—194, Beograd.

Dimitrijević, M. D. & *Dimitrijević*, M. N. (im Druck): Olistostrome melange in the Yugoslavian Dinarides and late Mesozoic plate tectonics. — The Journal of Geology, Chicago.

Gümüs, A. (1970): Türkiye metalojenisi. — Maden tetkik ve arama Enstitüsü yayinlarindan, Ankara.

Ivanov, R. (1966): The Rhodope Cenozoik petrographic province. — Referati VI. savetovanja geologa SFRJ, *II*, 94—114, Ohrid.

Ivanov, R. & *Stojanova*, Cv. (1966): Razsejani elementi v iztočnorodopskite vulkanski serii. — Trudove vrhu geologijata na Blgarija, Ser. geoh. min. i petr. BAN, *VI*, 83—102, Sofija.

Janković, S. (1967): Metalogenetske epohe i rudonosna područja u Jugoslaviji. — Rud. geol. fakultet i Rud. Institut, Beograd-Zemun.

Jovčev, J. (1965): Osnovi geologii i poleznie iskopaemie teritorii N. R. Bolgarii. — *VII* Kongres Karp. balk. geol. asoc., Sofija.

Karamata, S. (1962): Der tertiäre Magmatismus in den Dinariden Jugoslawiens. Seine Phasen und die wichtigsten petrochemischen Charakteristiken. — Referati V. Savetovanja geologa FNRJ, *II*, 137—148, Beograd.

Karamata, S., Knežević, V., Antonijević, I., Djordjević, M., Mićić, I., Divljan, M. & *Drovenik, M.* (1967): Les roches magmatiques crétacées-tertiaires des Carpato-Balkanides Yougoslaves. — Acta Geol. A. S. Hung., *XI/1*—3, 115—138, Budapest.
Karamata, S. (1969): Lead in sanidines from quartz-latites from the Zvečan ares, Yugoslavia. — Vesnik Zavoda za geol. i geof. istraživanja, A, *XXVII*, 267—276, Beograd.
Karamata, S. (1973): Petrologic, geochemical and metallogenetic provinces of Cretaceous and Tertiary age in Yugoslavia and neighbouring areas. — Colloque E. Raguin, Paris.
Karamata, S. (im Druck): Geochemical, pretrologic and metallogenetic provinces on the Balkan Peninsula and in Asia Minor. — Posebna izdanja SANU, Beograd.
Kolčeva, K. (1969/70): Vrhu petrohimijata na černozemsko-nazdelskia pluton. — Godišnik na Sof. Universitet, Geol.-geogr. fak. *62/1*, Geologija 235—257, Sofia.
Köksoy, M. (1967): Dispersion of mercury and other ore elements from mineral deposits in Turkey. — Thesis. Dept. of geology, Imperial College of science and technology, London.
Maksimović, Z. & Terzić, M. (1965): Geohemija vulkanskih stena Rudničkih planina. — Referati I Simpozijuma iz geochemije, SGD, 221—242, Beograd.
Petraschek, W. E. (1955): Großtektonik und Erzverteilung im mediterranen Kettensystem. — Sitz. Ber. österr. Akad. Wiss., math.-nat. Kl. *164*, Wien.
Petrascheck, W. E. (1969): Ore metals from the crust or mantle. — Economic geology, *64/5*, 576—578.
Ronov, A. B. & Yaroshevsky, A. A. (1969): Chemical composition of the Earth's Crust. — The Earth's Crust and Upper Mantle, Geophysical Monograph, *13*, 37—57, Washington.
Turekian, K. K. & Wedepohl, K. H. (1961): Distribution of the elements in some major units of the Earth's crust. — Bull. geol. soc. America, *72*, 175—191.
Vakhrushev, V. A. (1971): Ore (sulphide) separations in rocks of the upper mantle and some problems of endogenic oreforming. — Intern. Geochemical Congress, Abstracts of Reports, *I*, 96—97, Moscow.
Vinogradov, A. P. (1962): Srednie soderžanija himičeskih elementov v glavnih tipah izverženih gornih porod Zemnoj kori. — Geohimija, 7, 555—571.
Wedepohl, K. H. (1956): Untersuchungen zur Geochemie des Bleis. — Geochimica and Cosmochimica Acta, *10*, 69—148.
Wedepohl, K. H. (1962): Beiträge zur Geochemie des Kupfers. — Geologische Rundschau, *52*, 492—504, Stuttgart.
Wedepohl, K. H. (1967): Geochemie. — Sammlung Göschen, B. *1224*, Berlin.
Willie, P. J. (1971): The dynamic Earth: Textbook in geosciences. — John Wiley & Sons, Inc., New York.

Mögliche Beziehungen zwischen der Verteilung der Zn-Halte und den Pb-/Zn-Vererzungen in der Trias der Draukalkalpen in Österreich

Ludwig *Kostelka* & Elisabeth *Niedermayr**

Zusammenfassung

Die Draukalkalpen wurden durch 19 Profile quer zur Streichrichtung des Gebirges beprobt und dabei keine prinzipiellen Unterschiede im Zinkgehalt der triadischen Schichtfolge von Ost nach West festgestellt.

Das Verhalten des Backgrounds könnte jedoch in den Gailtaler Alpen dahingehend aufgefaßt werden, daß Schwellen- und Becken-Zonen von Ost nach West abwechselten.

Es bleibt zweifelhaft, ob zwischen östlichen Gailtaler Alpen und Karawanken eine Schwellenzone bestand. Dies könnte jedoch geologisch erklären, warum die Tektonik die verminderten Gesteinsmächtigkeiten in diesem Bereich restlos überwunden hat, so daß das Eis die stark beanspruchten Teilschollen bis auf geringe Reste ausräumen konnte.

Eine Ost-West-Gliederung in Schwellen- und Becken-Zonen deutet sich an und wird nur durch weitere Beprobungen und Einbeziehung mehrerer Elemente in die Analytik bestätigt oder widerlegt werden können. Die Charakteristiken der verschiedenen Vererzungstypen würden demnach jeweils einem definierten Ablagerungsbereich entsprechen und nur aus einer synchronen Entstehung zwanglos zu erklären sein.

Abstract

"Relations between geochemical background and lead-zinc ore occurences in the limestone-Alps of the river Drau (Austria)"

66 mostly small deposits of lead and zinc ores are known in the Middle Triassic camgites of the Mesozoic sediments which form the limestone Alps north of the River Drau.

* Dr. Elisabeth *Niedermayr*, Thimiggasse 15/I, A-1180 Wien, Austria. — Hochschuldozent Dr. Ludwig *Kostelka*, Bleiberger Bergwerks-Union, Radetzkystraße 2, A-9010 Klagenfurt, Austria.

This limestone range extends for abt. 200 km in east-west direction. The north-south dimension of today is only 15 km in average. It was intended to get informations about the general distribution especially of lead and zinc in these limestone mountains by 19 geochemical sections cutting in north-south direction the whole system.

The results confirmed by higher concentrations of zinc in certain sediments are that there is a general change of the basemental contents depending from the facies of the rock respectively from the sedimentary environment.

In the areas where big concentrations of lead and zinc are known, i. e. at Bleiberg, the Zn-values of the background do not increase generally in the ore-bearing Ladinian and Carnian sediments; hence we state that there is no influence of ore-occurrences on the geochemical background generally. Ore-bearing in sediments depends partly on the general situation but it seems of decisive importance, that there must have been local phenomena which have no effect to the general geochemical background.

The studies will be completed as the Drau-limestone Alps represent a more or less autochtonous system of sediments which are partly extraordinary well studied by the mining industry and offering therefore good suppositions even for special investigations.

Einleitung

Die Diskussion über die Entstehung der Blei-Zink- bzw. Baryt- und Flußspatvererzungen in Sedimentgesteinen betrifft ein wissenschaftliches Problem mit wirtschaftlichen Aspekten ganz außerordentlicher Bedeutung.

Die bei der Diskussion der Genese aufgeworfenen Fragen führten zu einer Untersuchung aller Faktoren und damit auch zu der Erkenntnis, daß die Erklärung einer Lagerstätte ohne Berücksichtigung ihrer geologischen Umgebung nicht möglich sei.

Die vorliegende Arbeit stellt den ersten Versuch dar, das Verhalten des geochemischen backgrounds in den Draukalkalpen auf ihre ganze Länge, insbesondere auf Grund des Zinkgehaltes der Sedimente, zu untersuchen. In erster Linie sollte festgestellt werden, ob durch die Zn-Halte der triadischen Ablagerungen die Vermutungen bestätigt würden, daß die Erzführung von Ost nach West in jeweils höhere Schichtglieder aufsteigt.

Die Arbeiten sind nicht abgeschlossen, so daß die Ergebnisse als vorläufige Resultate anzusehen sind.

Die Geländearbeiten wurden von *Niedermayr* ausgeführt und insbesondere auf ihre genetische Aussage hin interpretiert. Diese Ergebnisse wird *Niedermayr* an anderer Stelle veröffentlichen. Die vorliegende Auswertung, vor allem auf das Untersuchungsziel der nach Westen in der Sequenz aufsteigenden Vererzung, stammt von *Kostelka*.

Geographische Lage

Die Kalkalpen südlich des Flusses Drau erstrecken sich von Abfaltersbach nahe der österreichischen Westgrenze bis über die Ostgrenze des Landes hinaus nach Jugoslawien auf eine Länge von mehr als 200 km.

Der östliche Abschnitt der Draukalkalpen wird als „Karawanken" bezeichnet, während der westliche Abschnitt unter dem Namen „Gailtaler Alpen" bekannt ist.

Geologische Kurzbeschreibung

Die Draukalkalpen bauen sich aus triadischen Gesteinen auf, die nicht verfrachtet wurden und daher eine nahezu autochthone Stellung einnehmen. Auf Grund der geologischen Position nahe der Grenze zwischen Nord- und Südalpen treten jedoch weitgehende Differenzierungen der einzelnen Gesteinshorizonte in streichender Richtung, besonders aber quer dazu, auf.

Aus geologischer Sicht können die Draukalkalpen in fünf Abschnitte unterteilt werden:

1. Der westlichste Abschnitt, die Lienzer Dolomiten, reichen vom Westende der gesamten Kalkserie bis zum Gailberg. In diesem Abschnitt treten charakteristische, besonders mächtige Breccienlagen in den mitteltriadischen Ablagerungen auf.

Hier finden sich auch Jura- und Kreideschichten in größerer Ausdehnung.

2. Die westlichen Gailtaler Alpen, die vom Gailberg bis zum Ostende des Weißensees reichen.

In diesem Abschnitt sind gut gebankte Kalke und Dolomite im Ladin und Karn besonders deutlich. Riffbildungen sind bekannt.

Die deutliche Bankung in großen Mächtigkeiten hat *van Bemmelen* seinerzeit bewogen, den Begriff der Jaukenkalke einzuführen, die sich vom Anis bis ins Nor erstrecken sollten. Diese Auffassung kann nicht bestätigt werden.

Die Gailtaler Alpen, insbesondere in ihrem östlichen Abschnitt, zeigen eine Einengungstektonik, wobei die Einengung im Meridian Bleiberg auf Grund der dort vorliegenden regionalen Verhältnisse nahezu 50 % beträgt.

3. Die östlichen Gailtaler Alpen vom Ostende des Weißensees bis Villach sind durch den Blei-/Zinkerzbergbau Bleiberg-Kreuth in diesem Abschnitt besonders gut studiert und bekannt geworden.

4. Die Karawanken, die südlich von Feistritz im Rosental ansetzen, gliedern sich geologisch in einen Nord- und einen Süd-Stamm. Diese zwei Einheiten sind durch die sogenannte Alpin-Dinarische Narbe voneinander getrennt und zeigen zum Teil makroskopisch erkennbar verschiedene Entwicklungen in den einzelnen Stufen der Trias.

Die Tektonik ist im östlichen Teil, insbesondere im Karawanken-Nordstamm, durch eine Nordbewegung gekennzeichnet, wobei die Trias jedoch nur wenige Kilometer über ihr Vorland hinweg bewegt wurde.

4a. Im Nordstamm der Karawanken sind eine Reihe von Pb-/Zn-Vererzungen bekannt (Windisch-Bleiberg, Eisenkappel und der in Jugoslawien derzeit produzierende Bergbau Mežica).

4b. Der Karawanken-Südstamm zeigt spurenförmige Vererzungen von Bleiglanz und Zinkblende, die aber nirgendwo Anlaß für einen Bergbaubetrieb gewesen sind.

Die geochemische Beprobung und deren Ergebnisse

Aufgesammelt wurden frische Gesteinsproben in 19 Profilen, woraus sich bei einer Gesamtlänge der Draukalkalpen von rd. 200 km ein mittlerer Profilabstand von 10 km ergibt. Die Richtung der Profile wurde quer auf das generelle Ost-West-Streichen in möglichster Annäherung an die Nord-Süd-Richtung über die triadische Gesteinsfolge hinweggelegt.

Abweichungen von der Nord-Süd-Richtung waren vielfach geländebedingt unumgänglich.

Der vorgesehene Probenabstand von 50 bis 100 m konnte bedingt durch Gelände- und Aufschlußverhältnisse — nur annähernd eingehalten werden.

Die Proben wurden auf ihren Zink- und teilweise auf ihren Bleigehalt im Labor der Werksdirektion Bleiberg untersucht und die Blei- und Zinkhalte zuerst mit Dithizon, später polarographisch bestimmt.

Die analytische Auswertung auf weitere Elemente ist nach der abgeschlossenen Modernisierung des analytischen Zentrallabors der BBU in Arnoldstein vorgesehen.

Eine Bestätigung der seinerzeit aufgestellten systematischen Gliederung der Vererzung, wonach diese in den Draukalkalpen von Ost nach West in der Sequenz aufsteigen sollte, konnte auf Grund der allgemeinen Zn-Konzentrationen nicht gefunden werden. (Kostelka, 1966).

Es ergibt sich aber die Möglichkeit, die Ergebnisse so zu interpretieren, daß die Draukalkalpen der Länge nach in zwei Schwellenzonen und drei Bereichen mit größeren Sedimentationsmächtigkeiten unterteilt gewesen sein könnten. Demnach wären die drei seinerzeit erkannten Vererzungstypen zwangsläufig verschiedenen Teilbecken bzw. gegenläufigen Schwellenbereichen zugeordnet und könnten daher wesentlich natürlicher und überzeugender erklärt werden (Abb. 1).

Die östliche Schwellzone ist zwischen Bleiberg und Windisch-Bleiberg zu denken. Sie hat die Sedimentationsräume der östlichen Gailtaler Alpen von denen der Karawanken getrennt. Dem östlichen Abhang bzw. den dort gelegenen Teilstrukturen würde der Vererzungstypus „Karawanken" entsprechen. Dem westlichen Abhang ein System von Riffen (?) und Lagunen, dem der Vererzungstypus „Bleiberg" zuzuordnen wäre.

Die westliche Schwellzone, etwa im Bereich des Kreuzberges, würde demnach die Sedimentationsräume der östlichen und der westlichen Gailtaler Alpen voneinander getrennt haben. Diese Schwelle hätte daher die Grenze zwischen dem

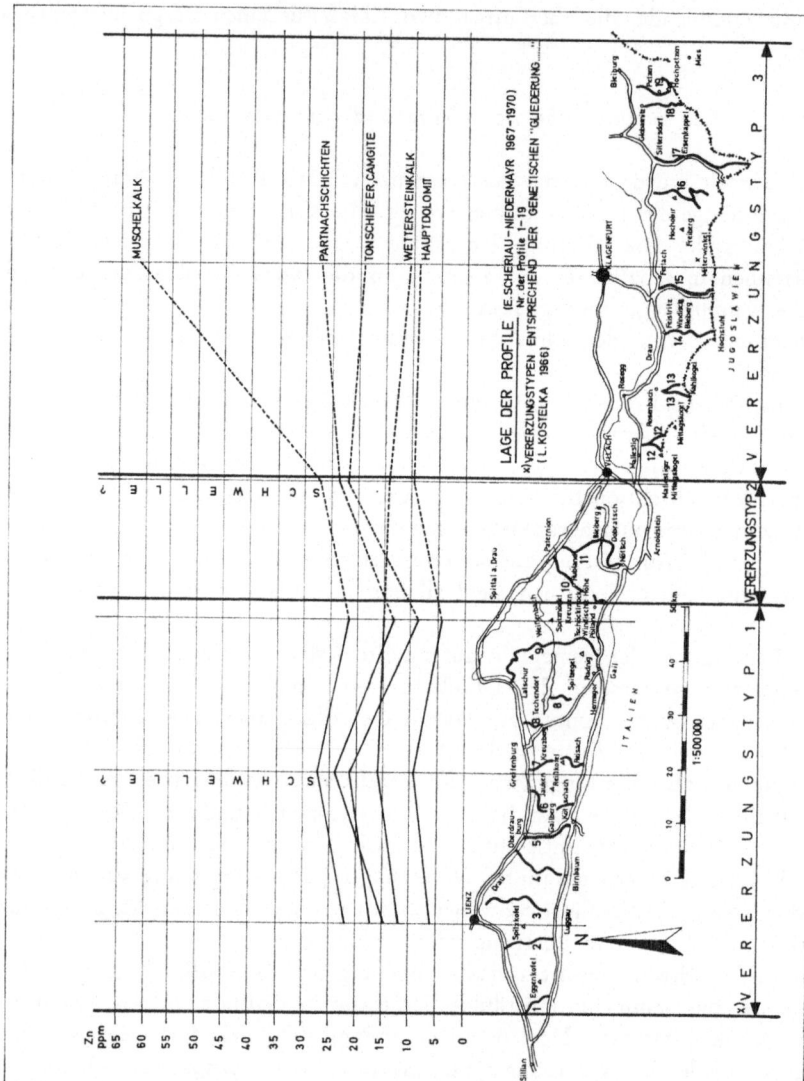

Abb. 1. Gegenüberstellung der Zn-Halte in den Drau-Kalkalpen.

Vererzungstypus „Bleiberg" im Osten und dem Typus „Jauken" im Westen dargestellt.

Unabhängig von diesen paläogeographischen Großgliederungen sind eine Reihe von örtlich wirksamen submarinen Relief- und Milieuunterschieden zum Teil bereits nachgewiesen worden, zum Teil kann dieser Nachweis noch erwartet werden.

Die Probenahme war auf die Ermittlung abgestellt, ob sich am geochemischen Hintergrund abzeichnet, daß eine Raum-Zeit-Abhängigkeit der für die Vererzung notwendigen Voraussetzungen im Bereich der Draukalkalpen insoferne bestanden hat als die Vererzung von Ost nach West in der Sequenz ansteigt. Es war zunächst nicht beabsichtigt, geochemische Aussagen zu machen, die über diese vermutete Trend-Ermittlung hinausgehen sollten.

Daß für das Auftreten der Blei-Zink-Vererzungen in den Draukalkalpen besondere, lokal begrenzte Verhältnisse maßgeblich waren, ist nicht nur von Bleiberg, sowie den Lagerstätten Mežica und Raibl anzunehmen, sondern gilt weltweit für diesen Lagerstättentypus. Dabei mag es sich um Faktoren gehandelt haben, die das Milieu weitgehend beeinflußten, die aber geologisch schwer oder überhaupt nicht nachweisbar sind (Meeresströmungen z. B.).

Aus diesem Grund ist nicht zu erwarten, daß das Verhalten des Zinkhaltes im allgemeinen (background), das sich am geochemischen Hintergrund abzeichnet, unbedingt mit dem Auftreten von Lagerstätten parallellaufen muß.

Tatsächlich ist das Auftreten von Lagerstätten in einzelnen Schichtgliedern der Trias nicht mit einer Erhöhung der background-Werte verknüpft. Dadurch wird die Annahme bekräftigt, daß regionale Voraussetzungen im Sinne *Tischendorfs* bei der Lagerstättenbildung eine Rolle spielen, daß aber lokale Verhältnisse von ganz entscheidender Bedeutung gewesen sind.

Hinweise auf einzelne Profile

Ein besonders charakteristisches Beispiel stellt das geochemische Profil Nr. 11 dar, das im Bereich der östlichen Gailtaler Alpen, und zwar im Meridian des Bergbaues Bleiberg von Süd nach Nord über die gesamte Einheit läuft. Interessant ist dieses Profil insbesondere im Vergleich mit den Ergebnissen der geochemischen Beprobung des sogenannten Rublandstollen, der auf 5 km Nord-Süd-Erstreckung den gleichen Abschnitt quert (Abb. 2).

Im Wettersteinkalk des Bleiberger Erzberges sind bei der Obertagebeprobung zwei hohe Werte von etwa 200 ppm Zn festzustellen.

Das charakteristische Verhalten der Zinkverteilung in diesem Bereich wird jedoch durch das Profil nicht wiedergegeben.

Dies liegt zum Teil an dem Umstand, daß schon von den Aufschlußverhältnissen her zwischen Untertage- und Obertageprobungen wesentliche Unterschiede bestehen.

126

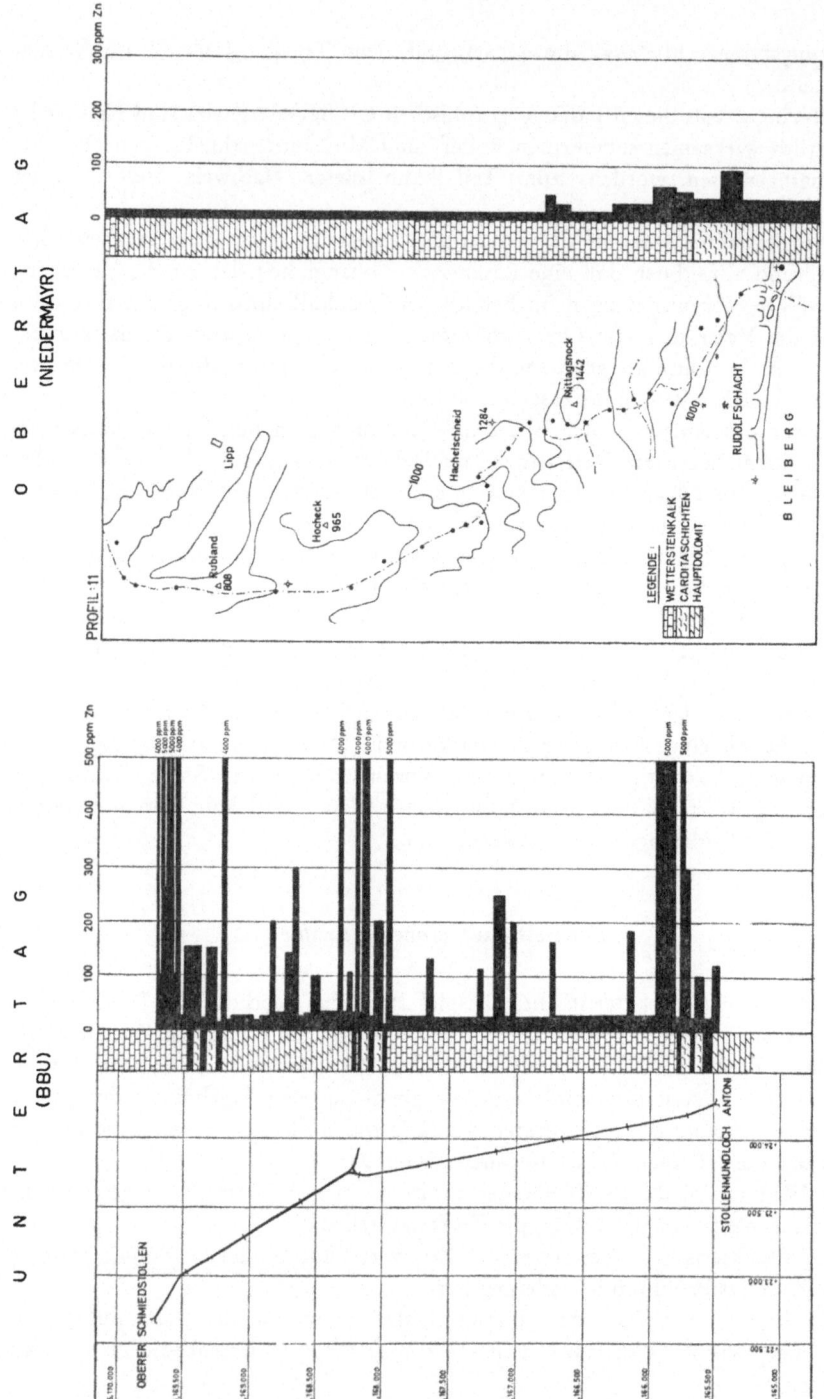

Abb. 2. Vergleich von Zn-Werten einer Übertage- und einer Untertage-Profillinie.

Das genau beprobte Untertagprofil des Rublandstollens zeigt zum Beispiel hohe Werte im hellen, obersten Wettersteinkalk der Rublandeinheit und in dem charakteristischen oberladinischen Kalk der Bleiberger Einheit. In beiden Bereichen sind bergwirtschaftliche Konzentrationen von Blei- und Zinksulfiden bekannt.

Auffallend im Untertageprofil sind engbegrenzte Maxima in beiden Flügeln der Rublandsynklinale in den karnischen Plattenkalken über dem 3. Tonschiefer in extrem gleicher stratigraphischer Position. In beiden Fundpunkten sind streng schichtgebundene Ober- und Untertage-Vererzungen bekannt, deren nähere Untersuchung derzeit anläuft.

Besonders charakteristisch sind jedoch in dem 5 km langen Rublandstollenprofil die Werte im Carditadolomit. In der Bleiberg-Einheit treten deutlich erhöhte Werte in dieser Stufe auf. In der makroskopisch gleich ausgebildeten Serie der Rublandeinheit jedoch sind keine erhöhten Zn-Halte festzustellen.

Es besteht kein Zweifel, daß die Vererzung dieses Dolomites in der Bleiberg-Einheit etwa 1,5 km westlich des Rublandstollens den Grund für dieses Verhalten darstellt. Es handelt sich dabei um die bedeutendste Konzentration von Blei- und Zinkerzen in der Lagerstätte Bleiberg.

Die äquivalenten Schichten in der Rublandeinheit sind erzfrei.

Diese Beobachtungen mahnen zur Vorsicht bei der Deutung und bei der Bewertung der Ergebnisse von Obertageprofilen in so kompliziert gebauten und stark differenzierten geologischen Bereichen. Dies gilt sowohl für die wirtschaftlichen Aspekte als auch für die Interpretation des geochemischen Grundverhaltens.

In den Profilen 5 bis 8 von *Niedermayr* sind in allen Schichtgliedern erhöhte Werte vorhanden. Dies stimmt gut mit der geologischen Vermutung überein, daß es sich insbesondere in den Bereichen, die vom Profil 8 und 9 gequert werden, um eine Schwellenzone handeln könnte, an deren westlichen Rand das Reißkofelriff zu liegen kommen würde.

Nach der Kartierung von *Bauer* in den Karawanken liegt im Bereich zwischen Petzen und Hochobir (Profil 19 bzw. 16) eine lokale Schwellenzone mit zum Teil stark verminderten Schichtmächtigkeiten vor. In diesen Profilen ist eine allgemeine Erhöhung der Zinkwerte, wie dies im Bereich vom Weißensee der Fall ist, (Profil 8) nicht festzustellen.

Die außerordentlich hohen Konzentrationen von Zink im Bereich des Hochobir-Kuhbergkammes insbesondere in der Karnischen Rekurrenz des Wettersteinkalkes sind bemerkenswert. Insbesondere deswegen, weil weiter östlich, im Bereich des Bergbaues Mežica, diese Rekurrenz in stratigraphische Einzelheiten geht, ohne daß größere und wirtschaftlich interessante Erzkonzentrationen festgestellt werden können.

Es ist beabsichtigt, die hier vorgelegten ersten Ergebnisse durch weitere Beprobungen zu ergänzen.

Geochemische und geochronologische Untersuchungen an sterilen und lagerstättenführenden Graniten insbesondere der Vendée (Frankreich)

F. Leutwein[*]

Zusammenfassung

Am Beispiel der petrographisch verschiedenen und verschieden entstandenen Granitmassive der Vendée wird gezeigt, daß granitogene Lagerstätten (Sn, Be, Li Mo, W, U) an bestimmte, gut bekannte Granittypen gebunden sind. Diese Gesteinstypen können sich auf sehr verschiedene Weise, magmatisch oder anatektisch, bilden. Die Granitisierung „in situ" kann gefolgt sein von magmatischen Mobilisierungen die zu einer Anreicherung leichtflüchtiger Stoffe führen kann, aber die große Menge von Granittypen kann in jedem Fall auf sehr verschiedene Weise entstehen — und nur die stets langdauernde Entwicklung von tiefplutonischen Biotitgraniten ± granodioritischen Chemismus über höhere Niveaus bis zu hochplutonischen, durch den Innendruck der leichtflüchtigen Substanzen mobilisierten Zweiglimmergranite, die dabei eine stets äußere Druckentlastung erfahren haben, führt zu Lagerstätten-höffigen Granitkuppeln. Die Dauer dieser Prozesse kann nach unseren Messungen 20—80 MA benötigen — insofern ist die Messung des Alters *eines* Granits wenig interessant. Die geochronologische Entwicklung innerhalb einer gegebenen Provinz granitischer Gesteine muß vielmehr ermittelt werden. Dabei hat die Untersuchung gewisser lithophiler Elemente und ihrer Verteilungskoeffizienten zu erfolgen. Hohe Sn-Gehalte in einem Biotitgranit besagen noch nichts über seine Lagerstätten-Höffigkeit. Nur wenn eine bestimmte Entwicklung oder Differenzierung dieses Granites zu jüngeren, in höheren Niveaus intrudierten Zweiglimmergraniten pegmatoiden Charakters (niederes K/Rb und Ba/Sr-Verhältnis) festgestellt werden kann, bestehen Aussichten auf lagerstätten-höffige Gebiete bei denen dann eine detaillierte geochemische Prospektion zu empfehlen wäre.

Abstract

Author studied the Vendée — the Southeastern end of the Armorican Massif in France. Here an important variety exists of different types of granitic rocks of different age and type of origin. The aim of this study is less to indicate

[*] Prof. Dr. F. *Leutwein*, Centre de Recherches Pétrographiques et Géochomiques, 15 Rue N. D. des Pauvers, F-54500 Vandoeuvre-lès-Nancy, France.

directly ore-deposits — our problem is to find in a rather important area of granitic rocks, those who might be excluded a priori from detailed prospection and those, which may be favorable for a further and more detailed prospection. The ore deposits or indices (Sn, Be, Li, Mo, W, O) are connected with granitic rocks and are always linked with certain well defined types of granite. But, we must take in mind, that nearly all types of granites may be the result of very different processus of origin — magmatic or anatectic. The granitisation in situ may be followed by magmatic mobilisations which may — or may not be accompanied by the enrichmet of volatile substances. Even if the chemical composition of major elements of the granites is identical, their origin may be different. Better informations are obtained by analyses on trace-elements. They allow better conclusions on their real history of evolution. Only the long lasting evolution from deep-plutonic biotite-granites of ± granodioritic character passing to higher levels of intrusive granites and finally to high plutonic (and even subvolcanics) diapiric intrusions of 2-mica-granites whose potassium-feldspar got often unstable in the last stages of evolution leads to the formation of cupolas which allow a favorable prognosis as regards ore-deposits. Our geochronological analyses showed us, that this processus may last 20 to 80 millions of years. So, the isolated dating of the age of one granite in a given province is not interesting. We have rather to follow by radiometric datations the historic evolution of the granitic rocks in the given province. Parallel to this, we have to study the evolution of certain lithophile elements (Rb, Sr, Li, Ba per ex.). High tin-contents alone are not significatives for the chances to find tin-deposits in such a granite. Only if a certain evolution or differentiation had been possible to younger, in higher levels intruded 2-mica-granites of pegmatoid character (low K/Rr and Ba/Sr ratio), we have real chances to find favorable metalliferous regions. And only under such conditions a detailed geochemical prospection could be recommanded.

Vor Jahren veröffentlichten wir eine geochemisch-metallogenetische Studie über die Zinnlagerstätten des Erzgebirges und ihre geochemische und tektonische Entwicklung (*Leutwein*, 1965). Die hier vorgelegte Arbeit bezieht sich auf Stadien der geochemischen Entwicklung, die vor der eigentlichen Lagerstättenbildung selbst liegen, sie entstand aus einer Untersuchung zur informativen Prospektion auf Uran und Zinn und anderen pneumatolytischen Erzlagerstätten. Dabei war es weniger wichtig, direkt Lagerstätten selbst zu finden als Regeln ermitteln, nach denen in sehr ausgedehnten Granitarealen einzelne Massive ausgeschlossen oder zurückgestellt werden konnten, die als weniger höffig oder als steril anzusehen waren. Es ist wohl bekannt, daß die meisten Zinn- und Uranlagerstätten zwar an Granite geknüpft sind, daß aber lange nicht alle Granite, obwohl vielfach petrographisch nahezu identisch, lagerstättenführend sind. Das gilt auch für die im allgemeinen günstigen Zweiglimmergranite. Als Arbeitsgebiet wählten wir insbesondere das armorikanische Massiv.

Für die in diesem Gebiet bekannten gangförmigen Pb-Zn-Sb-Lagerstätten varistischen Alters ergibt sich, daß sie fast ausnahmslos an die Nähe eines bedeutenden tektonischen Lineaments gebunden sind, das Lineament Brest-Alençon. Manche treten in Graniten oder in Granitnähe auf, andere nicht. Ein direkter Zusammenhang mit den Graniten verschiedenen Alters besteht nicht, wohl aber mit der tektonischen Situation. Zinn-Wolfram- und Uranlagerstätten sind dagegen überall strikt an bestimmte Granite gebunden und unabhängig von diesem Lineament.

Die Ergebnisse der geochronologischen Analysen

Die ältesten Granite Armoricas sind granodioritische, meist metamorphe Gesteine, die 2500 MA (Icartian bei Cherbourg, *Leutwein* et. al. 1973) bzw. ca. 900 MA (Pentévrien bei Erquy) alt sind und keine Lagerstätten führen. Etwas jünger sind die nicht metamorphen Granite der cadomischen Phase des jüngeren Praekambriums (Briovérien) im Bocage Normand. Zu ihnen gehört in der Gegend von Fougères, bei Montbelleux, die einzige nicht varistische Zinn-Wolframitlagerstätte Westeuropas, ca 500 MA alt — alle anderen sind erst im Lauf der varistischen Orogenese entstanden. Sie ist an einen diapirisch aufgedrungenen vergreisten Zweiglimmergranit gebunden (500 MA alt) der am Rande des großen, postorogenen Plutons von ca. 570 MA liegt. Weitere geochronologische Angaben finden sich in *Leutwein-Sonet* (1968).

Im Südostsporn Armoricas, der Vendée, treten eine Reihe von Granitmassiven auf, in denen vor allem wichtige Uranlagerstätten liegen, Indizien von Sn, Mo, W und Be fehlen nicht (s. die folgende Figur).

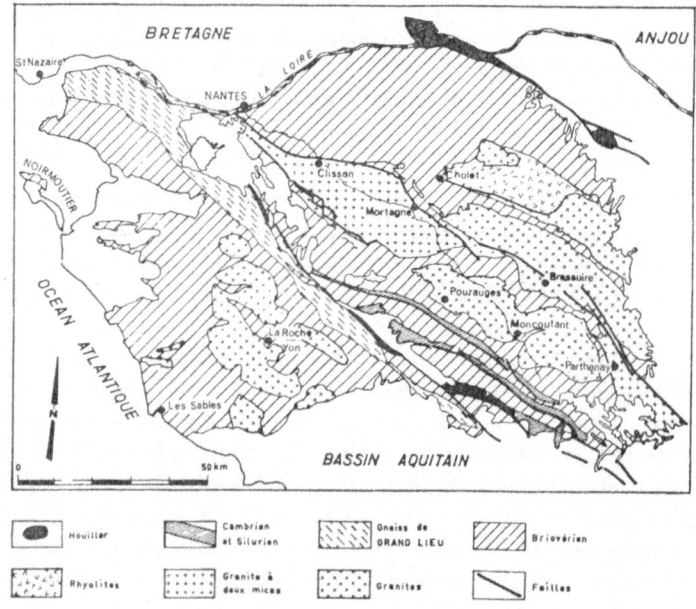

Fig. 1. Geologische Skizze des granitischen Massivs der Vendée.

Wir haben es mit fünf gut begrenzten Granitmassiven zu tun (Mortagne, Bressuire, Pouzauges, Neuvy-Bouin und Parthenay) in denen jeweils sehr ähnliche Gesteinstypen auftreten, von Biotitgraniten (± porphyrischer Struktur) bis zu Zweiglimmergraniten und selbst reinen Muscovitgraniten. Pegmatite fehlen nicht, doch sind die wirklich wichtigen Uranlagerstätten nur aus dem Granitmassiv der Mortagne bekannt. Es wurden geochemische, petrographische und geochronologische Methoden kombiniert angewendet. Über die analytischen Ergebnisse unterrichtet Tabelle 1.

Tabelle 1
Granite der Vendée

GRANITES de VENDÉE
(Teneurs en ppm sauf Na)

GRANITE der VENDÉE
(Gehalte in ppm soweit nicht anders angegeben)

LOCALITES		Min.	Pb	Sn	Be	Ba	Sr	Li	Na %	K %	Rb	K/Na	K/Rb	Rb/Sr	Ba/Sr	Ba/L.
(204)	Massif Pouzauges	R.T.	7	tr	-	500	158	20	2,34	2,91	75	1,24	390	0,48	3,1	25
	Granite à Biotite	Or	4	tr	-	5000	130	/	2,36	8,86	115	3,75	770	0,9	38	nd
		Bi	15	tr	15	2000	30	110	0,25	5,51	480	22,0	115	10	67	80
(224)	Granite à Amphibole	R.T.	9	tr	-	600	85	10	3,22	1,48	40	0,46	370	0,02	7,1	60,60
		Bi	-	tr	25	1500	67	50	0,60	3,71	120	6,2	309	1,8	22	30
(201)	Granite à 2 Micas	R.T.	25	15	-	tr	40	60	3,86	2,58	620	0,67	420	15,5	nd	nd
		Or	40	/	20	280	58	/	2,34	9,10	1600	3,9	57	27,6	4,8	nd
		Musc.	/	/	60	110	5	1060	0,68	8,25	4400	12,13	18,8	880	22	0,1
(318)	Massif Bressuire	R.T.	10	-	4	900	420	90	2,80	3,60	90	1,30	400	0,21	2,1	10
	Granite à 2 Micas	Or	80	-	-	4000	550	tr	1,70	10,20	320	6,0	319	0,58	7,3	nd
		Biot.	-	-	20	1500	17	850	0,25	7,17	880	28	82	52	88	2
(322)	Granite à Biotite	R.T.	5	-	10	800	237	70	2,90	3,73	190	1,3	196	0,8	3,4	11,4
	(Moulins)	Or	30	-	3	1500	338	10	3,70	6,05	260	1,6	233	0,8	4,4	150
		Bi	70	-	25	500	35	1200	0,40	4,31	600	10,8	72	17,2	14	0,4
(36)	Mortagne	R.T.	5	20	7	tr		290	2,80	3,50	560	1,25	62,5	22	nd	
	Granite à 2 Micas	Or	70	15	10	140	35	30	1,94	10,30	1140	5,3	90	32	4	4,7
		B	12	200	20	10	9	3300	0,32	7,02	2800	20,6	25	311	1	0,003
		M	-	240	40	tr	7	1100	0,92	7,91	1900	8,6	42	271	nd	
(60)	Granite à grain fin	R.T.	20	20	11	440	81	230	2,24	4,39	415	2,0	106	5,1	5,4	2
	et à 2 Micas	Or	17	tr	-	1100	149	20	1,51	10,97	717	7,3	153	4,8	7,4	55
		B	-	70	10	200	9,2	1970	0,30	7,20	1675	24,0	43	182	22	0,1
		M	-	200	100	400	19,2	750	0,74	7,62	863	10,3	88	447	21	0,5
(106)	Commanderie	R.T.	10	50	25	200	38	430	2,42	4,13	653	1,7	63	17,2	5,3	0,47
	Granite à 2 Micas	Or	50	-	-	750	74	50	1,30	11,12	1246	8,6	89	16,8	10	15
		B	-	300	25	90	9,2	4900	0,22	7,25	3840	33,0	19	417	10	0,02
		M	-	300	60	130	7	1940	0,60	8,22	1850	13,7	44	264	19	0,07
(2)	Granite porphyrolte	R.T.	20	10	7	540	104	130	2,48	3,94	300	1,6	131	2,9	5,2	4,2
		Or	55	-	-	1500	183		1,48	11,04	610	7,3	181	3,3	8	
		B	-	60	15	370	15	1180	0,40	7,14	1440	18,0	50	96	25	0,3
(93)	Granite à Muscovite	R.T.	5	50	10	300	37	300	2,34	3,77	575	1,6	66	15,5	8,1	1
		Or	25	tr	-	700		50	1,46	10,34		7,1				14
		M	-	240	60	210		1500	0,64	8,09		12,6				0:14
(401)	Parthenay	R.T.	tr	tr	7	240	50	50	3,61	5,14	325		132		4,7	4,8
	Granite à 2 Micas	Or	17	tr	4	700	90				535					
		B	30	50	-	tr	20			2,5	350					
		M	3	110	10	200	25				685					
(407)	Parthenay	R.T.		10	7	60	25	nd	3,60	4,64	380		69		2,4	
	Granite à 2 Micas	Or	11	tr	tr	150	55				990					
		B	tr	60		tr	8				2680					
		M	12	180	15	100	4				1640					
(414)	Parthenay	R.T.		tr		340	123		3,27	3,21	186		175			
	Granite à Biotite	Or	6			1400	225				330					
		B		20			8				820					
(853)	Guerande	R.T.	20	5	6	420	165	nd	3,96	4,60	220		174		2,5	
	Granite à 2 Micas	Or	35	tr	4	800	245				340					
		B	tr		5	18	25				1059					
		M	3	50	10	400	20				555					
(856)	Guérande	R.T.	25	/	8	350	188		3,80	4,60	229		172			
	Granite à Biotite	O	65	/	3	1300	307				465					
		B		10	/	nd	10			6,7	1181					
(866)	Guérande	R.T	10	/	12	300	122		3,77	4,61	260		147			
	Granite à Muscov.	O	35	/	5	550	233				375					
		M	3	50	15	280	19				644					

Bemerkt sei, daß die Gehalte an Pb, Sn, und Be uv-spektrographisch mit ca. ± 15 % mittleren Fehlern ermittelt wurden. Ba, Na, K und Li wurden flammen-photometrisch mit ca. 2 % mittleren Fehlern bestimmt. Die Werte für Rb und Sr folgten aus den massenspektrometrischen Analysen mit etwas höherer Genauigkeit (Isotopes-Dilution Technik). In petrographischer und geochemisch-metallogene-tischer Hinsicht ist das Massiv von Mortagne von J. *Renard* (1971) eingehend untersucht worden. Geochronologisch wurden die Proben nach der Rb/Sr und K/A-Methode untersucht, sowohl als Isochronen über Gestein + Mineralien wie auch als Gesamtgesteins-Isochronen. J. *Sonet* hat (1967 und 1968) zwei kurze Be-richte darüber veröffentlicht auf die hier verwiesen sei. Zusammenfassend ergab sich folgendes: K/A-Alter an frischen Glimmern und Feldspäten, bei feinkörnigen Gesteinen sogar an „Gesamtgesteinsproben", ergaben brauchbare Hinweise für die jüngsten Prozesse thermischer Art, die sich in dem Gebiet abgespielt hatten. Als Mittel von ca. 30 Datierungen über *alle* Granitmassive — sterile und lager-stättenführende — ergab sich 280 ± 10 MA, d. h. dasselbe Alter, das *Kosztolanyi* (1970) für die ältesten Pechblenden fand.

Daß die geologische Vorgeschichte dieser Massive sehr verschieden ist und daß ihre Differenzierung lange Zeiträume erforderte ist aus den K/A-Datierungen nicht ersichtlich. Aus den Rb/Sr-Messungen erfolgte zunächst, daß nur zwei Mas-sive in isotopischer Hinsicht homogen, bzw. homogenisiert sind, was sich dadurch äußert, daß Mineral-Isochronen und Gesamtgesteinsisochronen gleiche Ergebnisse liefern: Mortagne 300 ± 10 MA mit $Sr_i = 0,709$, Neuvy-Bouin das gleiche Alter, aber mit $Sr_i = 0,713$. Die der anderen Massive ergaben keine Isochronen sondern nur stark streuende Werte. Nur für Pouzauges ergab sich ein eindeutiger Trend für ein prae-varistisches Alter — 430 ± 30 MA.

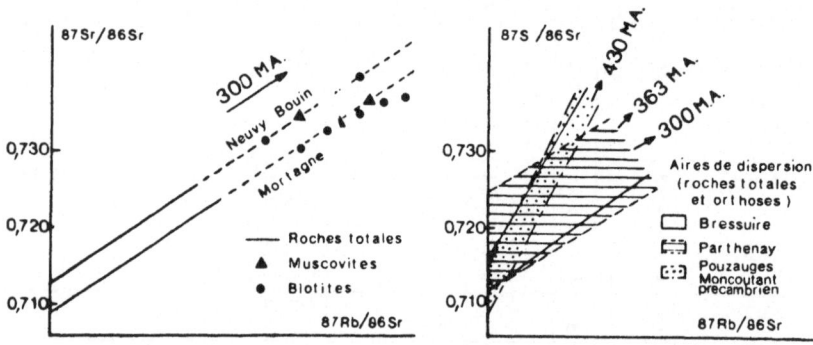

Fig. 2. Schematisches geochronologisches Diagramm (J. *Sonet,* 1968).

Auch der Versuch, alle Gesamtgesteinswerte der verschiedenen Massive in einer Isochrone zusammenzufassen scheiterte. Fig. 3 zeigt das Gesamtgesteinsdiagramm — allein ein allgemeiner Trend für varistische Alter läßt sich feststellen.

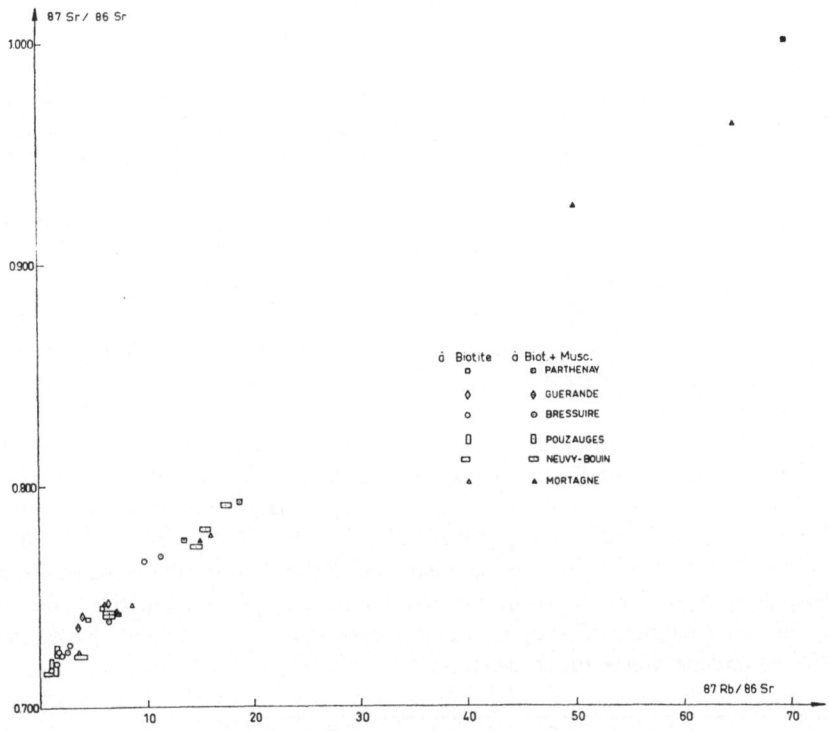

Fig. 3. Zusammenfassung aller Gesamtgesteinsanalysen im Nicolaysen-Diagramm.

Die Werte einzelner Massive könnte man zwar zu speziellen Isochronen zusammenfassen, aber es ist völlig klar, daß die von *Nicolaysen* aufgestellten Voraussetzungen für die Anwendbarkeit der „Gesamtgesteinsmethode" hier nicht erfüllt sind. Dafür ist Alter und Geschichte, isotopische Zusammensetzung und geochemische Entwicklung der verschiedenen Massive zu verschieden. Wichtig ist festzuhalten, daß nur das Massiv von Mortagne als einziges, und randliche Teile der Massive von Parthenay und Moncoutant lokal Gesteine darstellen, in denen der Entwicklungsvorgang bis zu fast schon pegmatoiden bzw. Greisentypen durchlaufen wurde. (Homogene junge Rb-Sr-Alter aller Mineralien und übereinstimmend mit den K/A-Altern.) Stratigraphisch gesehen gehören diese Gesteine und ihre Lagerstätten an die Wende Stephanien/Autunien und jünger. Das Ganze möge als Beispiel dienen, zu zeigen, wieviel geologisch-petrographische Kriterien berücksichtigt werden müssen, um die „Total-rock-isochrone" Methode auf ausgehende Granitareale sinnvoll anwenden zu können.

Die Ergebnisse der geochemischen Analysen

Eine Erzlagerstätte stellt stets eine Zone einer geochemischen Anomalie dar, in der bestimmte Elemente erheblich über ihren Clarke-Wert hinaus angereichert sind. Für einige Metalle wie z. B. Sn, U, betragen die erforderlichen Anreicherungsfaktoren 1000 und mehr, um auch nur annähernd zu den heute abbaufähigen Mineralgehalten zu gelangen. Derartige Metallanreicherungen sind auf dem Wege einfacher, direkter chemischer Reaktion nicht erhältlich. Es müssen sich bestimmte Umlagerungsprozesse in mehrfacher Folge abgespielt haben um von einem Granitmassiv mit z. B. 3 g Sn/t zu einer Zinnlagerstätte mit ca. 5000 g/t zu führen. Dabei ergibt eine Nachrechnung von Erzlagerstätten bekannten Gehalts und bekannter Ausdehnung, daß der Wirkungsgrad dieser Anreicherungsprozesse selbst gering ist und um so geringer wird, je höher der Anreicherungsvorgang war.

Am Beispiel der gut bekannten Sn-Lagerstätten des Erzgebirges läßt sich zeigen, daß in den aus der Tiefe aufgestiegenen Diapir-Graniten, die hochplutonische Niveaus erreichten, viele tausend ton. Sn vorhanden sind. Der Metallinhalt in den viel selteneren Reichst-Erz-Zonen mit mehr als 5 %/o Sn ist demgegenüber fast unbedeutend. Fig. 4 gibt ein schematisches Beispiel eines diapirischen Zinngranits, der aus größerer Tiefe in ein hochplutonisches Niveau aufstieg. Die Dimensionen sind häufigen Vorkommen angenähert, den Gehaltsangaben liegen tatsächlich gefundene Werte zugrunde.

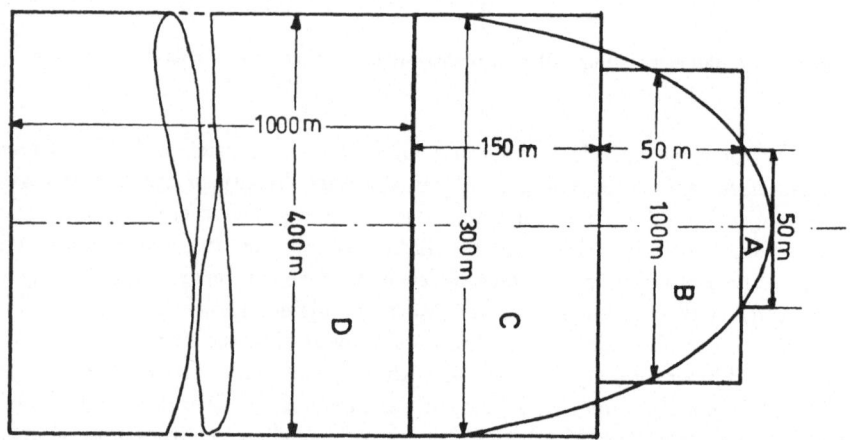

Fig. 4. Metallgehalte und Metallinhalte in verschiedenen Zonen einer Zinngranitkuppel.

Über Ionenradien und die Clarke-Werte der verschiedenen geochemisch wichtigen Alkali- und Erdalkalimetalle unterrichtet die folgende Tabelle, in der auch einige geochemisch wichtige K/Rb- und Ba/Sr-Coeffizienten angegeben wurden.

Tabelle 2
Clarke-Werte wichtiger Gesteinsgruppen (in g/t = ppm soweit nicht anders angegeben) *(Turekian & Wedepohl, 1961)*

C l a r k e s — Granite

Elem.	I. rad.	Basalte	> Ca	< Ca	Syenite	Schiefer	Sandst.	Tone
Ba	1,44	330	420	840	1600	580	10	2300
Sr	1,21	465	440	100	200	300	20	180
Cs	1,78	1,1	2	4	0,6	5	0,0×	6
Rb	1,57	30	110	170	110	140	60	110
K	1,46	8300	2,52 %	4,20 %	4,80 %	2,60 %	1,07 %	2,50 %
Li	0,82	17	24	40	28	66	15	57
Ba/Sr		0,7	0,95	8,4	8	1,9	0,5	12,7
K/Rb		280	230	250	430	190	180	230
Ba/Li		19,4	17,5	22,0	57,1	8,7	0,66	40,3

Unter Entwicklung der Granite, insbesondere der Kalkalkalireihe, verstehen wir den Übergang von Granodioriten, die in situ granitisierte Volumelemente der Erdkruste blieben, oder die nach der Granitisierung als Magmen intrusiv wurden. Homogen sind solche Gesteine fast nie. Der Chemismus ist granodioritisch oft mit Natrontendenz. Bei weiterer Differenzierung — die nicht unbedingt stattgefunden haben muß, entstehen Biotitgranite, oft mit späteren Kalifeldspatporphyroblasten. Die Entwicklung geht weiter zu Zweiglimmergranit oder Muscovitgraniten die meist gleichkörnig, manchmal, in subeffusiven Typen sogar miarolitisch sein können. Die Entwicklung läßt sich an Hand der zunehmenden SiO_2-Werte verfolgen. Auch der Natrongehalt der Plagioklase nimmt in den entwickelten Fazies zu, bis zu Gesteinen mit überwiegend Albit + Lithiumglimmern. Natron- und Kalitendenzen treten nacheinander manchmal im Wechsel auf. Zuletzt ist Orthoklas instabil. Eigentliche Alaskite sind nur sehr bedingt als günstig zu bezeichnen — vielfach enthielten diese extrem sauren Magmen zu wenig leichtflüchtige Substanzen um eine Anreicherung an Erzmineralien zu ermöglichen. Ihr Ba/Sr-Koeffizient bleibt dann hoch (über 3) trotz des relativ kleinen K/Rb-Koeffizienten.

Dazu ist eine kristallographisch-geochemische Bemerkung notwendig. Die wichtigsten Mineralien granitischer Gesteine sind Kalifeldspat, Plagioklas, Biotit und Muskovit. Über die Verteilung der Alkalien und Erdalkalien ist zu bemerken, daß Ba und Rb im allgemeinen dem Kalium folgen. Sr kann zwar isomorph das Kaliumion ersetzen, folgt aber im allgemeinen eher dem Calcium — findet sich also vor allem in den Plagioklasen, weniger in den jüngeren Albit-Neubildungen (s. Tab. 3).

Sr wird in Glimmern nur als Spurenelement zugelassen, in Plagioklasen tritt es aber als kristallochemisch wichtige Komponente auf. Auch Chlorite können größere Mengen Sr einbauen. Ba z. B. ist trotz seiner Zweiwertigkeit bevorzugt in Kalifeldspäten eingebaut (das kann bis zu 1 % Celsianmolekül in Orthoklas oder Mikroklin gehen, vor allem in späteren Kalifeldspat-Porphyroblasten). Im allge-

Tabelle 3
Verteilung der seltenen Alkalien und Erdalkalien in granitbildenden Mineralien

	Ba	Sr	K	Na	Li	Sn
Orthoklas	+ +	Sp.	+ +	—	Sp.	—
Plagioklas	—	+ +	—	+	—	—
Muskovit	+	Sp.	+ +	—	+ +	+
Biotit	+	Sp.	+ +	—	+ +	+ +

meinen, s. Tab. 3, enthält Orthoklas etwa halb soviel Rb wie der Biotit des gleichen Gesteins — Kalifeldspäte mit über 0,1 % Rb sind offenbar nicht stabil. Auch Glimmer können beträchtliche Mengen Ba speichern, nicht aber die Plagioklase. Der Koeffizient K/Rb stellt also eine Beziehung zwischen dem nicht eben leichtflüchtigen K und dem — zusammen mit Cs durchaus mobilen Rb dar — ohne Berücksichtigung der Plagioklase. Der Koeffizient Ba/Sr dagegen, der zwei nicht flüchtige Elemente zusammenfaßt, berücksichtigt vor allem das Kalifeldspat + Glimmerverhältnis zu den Plagioklasen und schließt die als Sammler von leichtflüchtigen Elementen (Li, Sn) wichtigen Glimmer praktisch aus. Wichtig ist er vor allem in granitischen Gesteinen der Kalk-Alkalireihe vom Granodiorit bis zum Zweiglimmergranit. Mit dem Auftreten von Mikrolin-Muskovitpegmatiten und bei Vergneisungsprozessen ergeben sich sehr streuende Werte. Im Sinne von *Jedwab* (1953) sind Rb und Ba kristallochemisch an K fixierte Elemente, Sr ist kristallochemisch an Plagioklase gebunden, in Glimmern ist es mehr ein typochemisches Spurenelement. Bei der Auswertung der K/Rb-Werte folgen wir vor allem D. M. *Shaw* (1968). Das meiste Sr das wir in Glimmern finden ist also normalerweise ein rubidigenes 87 Sr. In Plagioklasen wie in Apatiten überwiegt dagegen das „normale" Sr. Im Laufe der oben geschilderten Entwicklung granitischer Gesteine nimmt der absolute Gehalt an Ba ständig ab, Sr bleibt eher konstant. Damit verringert sich der Koeffizient von hohen Werten (3—8) im Laufe der Entwicklung zu lagerstättenhöffigen Differenziaten bis an Werte von 0,8—2. In granitisierten Arkosen z. B. bleibt er hoch, auch führen saure, aplitische Leukosome die ev. als selbständige Leukogranite, oft gneissiger Struktur, schon zu Anfang eines orogenen Cyklus auftreten können, höhere Ba/Sr-Werte, selbst wenn ihr Rb/Sr-Verhältnis niedere, scheinbar pegmatitische Werte annehmen sollte. Solche Granite sind in metallogenetischer Hinsicht uninteressant. Manche Alaskite gehören hierher, viele Teile der „granulites feuilletées" des Trégorrois und als bestes Beispiel der früh varistische Klemmbachgranit des Schwarzwaldes (n. *Brewer* & al., 1972, 340 MA). Selbst Muscovit-Pegmatite treten in diesem Gestein auf, nur ist es völlig steril. Selbst Spuren von Sn waren nicht nachweisbar, doch enthält es ca 20 ppm Pb, sein K/Rb-Wert ist 170, also schon nahe den Werten pegmatoider Gesteine. Ba/Sr bleibt mit 3,3 aber noch im Gebiet wenig differenzierter Granite, trotz seines Charakters als saurer Zweiglimmergranit.

Daß ein und derselbe Granittyp auf sehr verschiedene Weise entstehen kann ist längst bekannt. Aber nur ein bestimmter Typ von Zweiglimmergraniten, der am Ende einer langen, nach unseren Beobachtungen leicht 20 bis 80 Millionen Jahren dauernden Entwicklung steht, erweist sich immer wieder als günstig für weitere Prospektionsarbeiten. Nur er kann lokal luxullianitische Züge — oder aber auch aus Quarz, Albit, Lithionglimmer bestehende Gesteinstypen enthalten. Hier ist vor allem auf die eingehende Veröffentlichung von *Lameyre* (1966) hinzuweisen, sowie auf die Arbeiten von *Ilyama* & al. (1963). Immer müssen bestimmte magmatisch-deuterische Entwicklungsstufen durchlaufen werden um zu der notwendigen Metallanreicherung zu führen. Die Migration von Ba, Sn, Li, Be, offenbar auch W, Mo, ist unabhängig vom Oxydoreduktionspotential, leicht aber durch Auftreten einer endogenen Anatexis auszulösen. Die Anreicherung von U wird dagegen entscheidend beeinflußt von dem wechselnden Oxydo-Reduktionspotential. Das erklärt auch warum in den Anfangsstadien U und Sn ziemlich parallel gehen, sich aber (z. B. bei der deuterischen Chloritisierung der Glimmer) grundsätzlich trennen. Während die Früh-Ausscheidungen von winzigen Uraninitkristallen stets beträchtliche Mengen von Th enthalten, sinkt im Laufe der anatektischen, bzw. deuterischen Entwicklung der Th-Gehalt immer mehr bis schließlich die abbauwürdigen U-Lagerstätten nur Th-freie Pechblende enthalten (s. z. B. *Renard*, 1971). Die Migration von Th IV ist abhängig von der Alkalinität der migrierenden Stoffe, aber unabhängig vom Oxydo-Reduktionspotential. In den diapirischen, jüngeren Zweiglimmergraniten mit pneumatolytischer Nachphase ist Uran nicht mehr beteiligt, doch treten besonders hydrothermale Lagerstätten dieses Metalls gelegentlich auch dort auf, wo eine solche saure und oxydierende Phase nicht durchlaufen, d. h. wo die Glimmer nicht chloritisiert wurden. Obligatorisch ist aber mindestens das Durchlaufen verschiedener granitischer intrusiver Fazies bis zu den Zweiglimmergraniten die dann oft chloritisiert sind. Aus diesen Überlegungen folgt, daß die „potentielle" Lagerstättenhöffigkeit an Hand von Analysen einiger Alkalien und Erdalkalien untersucht werden kann. Offen bleibt natürlich die Frage, ob denn ein „Angebot" an dem betreffenden Metall überhaupt vorlag. — Die analytische Verfolgung der Sn, Be, Li-Gehalte etc. ist zwar notwendig, doch ist sie allein unzureichend zur Interpretation. Bei dem aus diesen Argumenten als „sehr günstig" zu beurteilendem Massiv der Mortagne war das „Angebot" gegeben, im Massiv von Parthenay, trotz günstiger Indizien, weniger.

Ein ziemlich einfaches geochronologisches Indiz, ob wir es mit banalem Granit zu tun haben, der nach seiner „mise en place" eine weitere Entwicklung nicht durchlaufen hat, besteht in der Datierung der Glimmer nach der Rb/Sr-Methode (bei Glimmern ist die genaue Kenntnis der Sr_i-Werte für eine Datierung nicht von so entscheidender Bedeutung) und gleichzeitig die Datierungen der Kalifeldspäte nach der K/Ar-Methode. Sind die Alterswerte sehr verschieden, so ist eine deuterische (oder tektonische) Überprägung des Granits sehr wahrscheinlich und damit die Chancen der Lagerstättenhöffigkeit gering. Erst bei völliger Aufschmelzung und Differenzierung zu Zweiglimmergraniten werden die Perspektiven günstiger. Wichtig wird damit vor allem die systematische Untersuchung des K/Rb- und Ba/Sr-Wertes der Gesteinsproben. Dazu kommen Analysen auf die interessierten Metalle. Gerade für diese Analysen genügen schnelle, preiswerte Analysenverfahren ohne allzu große Präzision.

Danksagung

Ich möchte hier vor allem Herrn Prof. M. *Roubault,* dem Direktor unseres Institutes für seine Hilfe und Förderung danken. Ebenso geht mein Dank an die Kollegen der „Commission à l'Energie Atomique", die diese Arbeit angeregt haben, und die auch das Untersuchungsmaterial nach unseren Angaben und vor allem in gutem Erhaltungszustand zur Verfügung stellten. Alle Proben wurden durch Sprengung frischen Materials gewonnen — ein für geochronologische Untersuchungen besonders wichtiger Punkt. Besonders gilt dieser Dank den Herren *Gangloff, Moreau* und *Gerstner.* Im Institut in Nancy bin ich allen meinen Mitarbeitern zu Dank verpflichtet, vor allem Herrn J. *Sonet.*

Literaturverzeichnis

Brewer, H. S. & *Lippolt,* H. J. (1972)- Rubidium-Strontium-Altersbeziehung variszischer Granite des südlichen Schwarzwaldes. — Fortschritte Min., *50,* Beiheft 3, p. 5—6.
Iiyama, J. T., *Wyart,* J. & *Sabatier,* G. (1963): Equilibre des feldspaths alcalins et des plagioclases, à 500, 600, 700 et 800^0 C sous une pression d'eau de 1000 bars. — C. R. Ac. Sci., *1963,* 256, p. 5016—5020.
Jedwab, J. (1953): Etude des oligo-éléments dans les minéraux des pegmatites. — Thèse, Bruxelles, *1953,* 159 pages.
Kosztolanyi, C. & *Coppens,* R. (1970): Etude géochronologique du gisement uranifère du Chardon (Vendée). — „Eclogae", vol. *63,* 185—196.
Lameyre, J. (1966): Leucogranites et muscovitisation dans le Massif Central Français. — Ann. de la Fac. Sci. de l'Univ. de Clermont, *12,* no 29.
Leutwein, F. (1965): Les caractéristiques géochimiques de l'évolution des gisements d'étain et de wolfram de l'Erzgebirge (Allemagne). — Sciences de la Terre, T. *X,* no 1, p. 35—78.
Leutwein, F., *Sonet,* J. & *Zimmermann,* J. L. (1968): Géochronologie et évolution orogénique précambrienne et hercynienne de la partie Nord-Est du Massif Armoricain. — Sciences de la Terre, Mémoire no *11,* 1—84.
Leutwein, F., *Power,* G., *Roach,* R. & *Sonet,* J. (1973): Quelques résultats géochronologiques obtenus sur des roches d'âge précambrien du Cotentin. — C. R. Acad. Sc. Paris, t. 276, série D, 2121—2124.
Nicolaysen, L. O. (1962): Graphic interpretation of discordant age measurements of metamorphic rocks. — Ann. Ac. Sci., New York, *91,* 148—206.
Renard, J. P. (1971): Etude pétrographique et géochimique des granites du district uranifère de Vendée. Liaisons entre l'évolution minéralogique et le comportement de l'uranium. Conséquences pour la prospection. — Thèse, Nancy, *1971.*
Shaw, D. M. (1968): A review of K/Rb-fractionation trends by covariance analysis. — Geoch. Cosm. Acta, vol. *32,* no 6, p. 573—601.
Sonet, J. (1967): Contribution à l'étude géochronologique du Massif de Mortagne (Vendée). — C. R. Acad. Sci, Paris, *264,* 225—228.

Sonet, J. (1968): Essai d'interprétation d'un ensemble de mesures géochronologiques au Rb/Sr des massifs granitiques vendéens. Mise en évidence d'une dualité génétique. — C. R. Acad. Sci. Paris, *267*, p. 15—17.

Tischendorf, G. (1969): Über die kausalen Beziehungen zwischen Granitoiden und endogenen Zinnlagerstätten. — Zeitschr. für angew. Geologie, *15*, H. 7, 333—342.

Tischendorf, G., *Lächelt*, S. & *Haake*, R. (1971): Zur Problematik der Verteilung von Zinnmineralisationen in Raum und Zeit. — Ber. deutsch. geol. Wiss., A, Geol. Paläont., *16*, 3, 379—394.

Turekian, K. K. & *Wedepohl*, K. H. (1961): Distribution of the elements in some Major Units of the Earth's Crust. — Geol. Suc. of Amer. Bull., *72*, 175—192.

Présentation d'une carte tectonique du monde, destinée à faciliter l'élaboration d'une stratégie de prospection des gîtes liés aux complexes basiques et ultrabasiques

Boris *Choubert*[*]

Abstract

Presentation of a tectonic map of continents at a scale of 1/10,000.000 based on the principle of tectonic stages and aimed at serving as a base for studies on the relationships existing between metalliferous deposits and the geological environment.

Zusammenfassung

Es wird eine Tektonische Karte der Erde im Maßstab 1/10,000.000 vorge-stellt, die nach der Methode der Phasengliederung zusammengesetzt ist und über die Beziehungen zwischen den Erzlagerstätten und ihrer geologischen Umgebung Aufschluß geben soll.

La répartition des matières utiles à la surface de la terre est loin d'être uni-forme et l'on a appelé "ceintures" ou "provinces métallogéniques" les zones privilégiées où se produisent des concentrations. La position géographique a donc une grande importance, et ceci nous a amené à dresser une carte permettant d'établir une relation entre la position des gîtes et les structures tectoniques du globe.

[*] Dr. B. *Choubert*, Résidence Bernard Palissy 10, F-77210 Avon, France.

Ce travail s'inscrit dans un ensemble de recherches entreprises par l'Ecole des Mines de Paris et animées par le Professeur P. Laffite: le projet *Basimines* tend à élaborer une stratégie de prospection des gîtes liés aux complexes basiques et ultrabasiques. Un progrès décisif sera fait dans ce domaine lorsqu'un recensement aussi complet que possible des connaissances acquises aura été réalisé à l'échelle mondiale et qu'un mode de traitement de cette masse de données aura été mis au point.

Deux stades doivent être envisagés:

1. Collecte et classement de l'information

— carte tectonique des continents à l'échelle du 1/10.000.000, exécutée suivant le principe des étages tectoniques et permettant le découpage des continents en unités structurales homogènes, en fonction de l'âge. Chacune de ces unités sera ensuite subdivisée d'une façon plus ou moins arbitraire en sous-unités représentant la surface d'un carré de 100×100 km, auxquelles seront rattachés les complexes et massifs de roches basiques et ultrabasiques, minéralisés ou non. Ces derniers se trouveront ainsi situés dans le temps et dans l'espace;

— constitution de plusieurs fichiers: position géographique, âge, géologie; description pétrographique et géochimique; métallogénie.

— un soin particulier sera apporté à noter la présence ou l'absence des divers métallotectes.

2. Exploitation de la documentation. Mise au point de modèle de simulation

— introduction de différents paramètres liés aux complexes proprement dits et aux régions naturelles du globe;

— étude de la répartition statistique des massifs divisés en complexes fertiles et stériles pour chaque type de minéralisation;

— mise en évidence, pour chaque unité structurale et chaque minéralisation, des métallotectes positifs et négatifs;

— mise au point d'une stratégie de prospection des gîtes d'un type donné, en intégrant dans des modèles de simulation les résultats obtenus précédemment.

Zur Minerogenie der Fluorit-Baryt-Lagerstätten in Mitteleuropa

Ludwig *Baumann* & Otto *Leeder*[*]

Zusammenfassung

Nach einem Überblick über die wichtigsten Fluorit-Baryt-Lagerstättenbezirke Mitteleuropas werden die gemeinsamen strukturellen und paragenetischen Merkmale hervorgehoben. Trotz lokaler Unterschiede läßt sich für die meisten Lagerstätten ein gemeinsames Standard-Paragenesenschema anwenden, das durch die Minerale Quarz, Baryt, Fluorit, Fe-Mn-Karbonate bzw. -Oxide, durch wechselnde Mengen an Sulfiden (Cu, Pb, Zn) und durch die Elemente Co, Ni, Bi, As, Sb, Ag und Hg charakterisiert ist. Die strukturelle Gemeinschaft ist die Bindung an herzynisch streichende saxonische Brüche, insbesondere an deren Kreuzungsbereichen mit rheinischen, eggischen und erzgebirgischen Strukturen. Untergeordnet treten auch synsedimentäre sowie tertiäre Bildungen auf. Die Lagerstätten lassen sich auf Grund der geotektonischen Situation, von physikalisch-chemischen Untersuchungen sowie geochemischen und petrogenetischen Überlegungen mit einem simatisch-juvenilen und atlantisch differenzierten Tiefenmagmatismus in Verbindung bringen.

Abstract

"About the minerogeny of the Fluorite-Barite-Deposits in Central Europe"

After a short review on the most important fluorite-barite deposits of central Europe, the common structural and paragenetic features are pointed out. In spite of local differences, there can be used one paragenetic model for all deposits, characterized by quartz, barite, fluorite, Fe-Mn-carbonates resp. -oxides, by varying enrichment of sulfides (Cu, Pb, Zn), and by the elements Co, Ni, Bi, As, Sb, Ag und Hg. The structural community is their binding to Hercynic striking

[*] Prof. Dr. habil. L. *Baumann* und Dr. O. *Leeder*, Sektion Geowissenschaften der Bergakademie, AG. Lagerstättenlehre und Minerogenie, DDR-92 Freiberg, DDR. Sektionsveröffentlichung Nr. 318.

Saxonic faults, particularly where they are crossing "Rheinische", "Eggische", or "Erzgebirgische" structures. Subordinate, there are synsedimentary as well as tertiary formations. The ore deposits can be connected with simatic-juvenile, and with differentiated "Atlantic"-type magmatism by their geotectonic situation, by physico-chemical investigations, as well as by geochemical and petrogenetic considerations.

1. Einführung

Der geologische Bau Mitteleuropas ist charakterisiert durch eine komplizierte Überlagerung und Verknüpfung mehrerer geologischer Epochen, die ihre tektonischen, magmatischen und metamorphen Spuren in diesem Gebiet hinterlassen haben. Diese geologischen Ereignisse haben an verschiedenen Orten und zu verschiedenen Zeiten Anlaß zur Konzentration von bestimmten Elementen geführt, die uns heute als Erz- bzw. Minerallagerstätten bekannt sind. Die meisten dieser Lagerstätten lassen sich nach dem gegenwärtigen Kenntnisstand und auf Grund einer umfangreichen wissenschaftlichen Bearbeitung in einen zeitlich und räumlich klar definierten minerogenetischen Rahmen stellen. Dabei ist eine weitgehende Klärung der Genese von Fluorit-Baryt-Lagerstätten, die in Mitteleuropa intensiv und extensiv weit verbreitet sind, erst in den letzten 15 Jahren möglich gewesen. Obwohl ein Blick auf die Verbreitung dieser Lagerstätten bereits eine regionale Häufung in bestimmten Gebieten erkennen läßt, wurden diese Vorkommen genetisch häufig unterschiedlich eingestuft. Die bisherigen genetischen Anschauungen sind verständlich, wenn man die räumlich z. T. weit auseinanderliegenden Einzelvorkommen oder Distrikte isoliert in ihrem engen geologischen Rahmen betrachtet und wenn man weiterhin berücksichtigt, daß die wissenschaftliche Bearbeitung meistens nicht unter einheitlichen Gesichtspunkten erfolgte. Mit wachsendem Kenntnisstand traten jedoch für die größte Anzahl dieser Lagerstätten weitgehende Gemeinsamkeiten in Erscheinung. Wenn auch lokale Besonderheiten, unterschiedliche Auffassungen der Bearbeiter und noch einige offene Probleme eine zusammenfassende Deutung erschweren, so sind doch bereits wichtige Fakten über eine weitgehende Analogie in struktureller, paragenetischer und genetischer Hinsicht gesammelt worden.

Das zu betrachtende Lagerstättengebiet Mitteleuropas wird begrenzt durch das Brahmsche Massiv im NW, den Rheintalgraben im W, den Alpenbereich im S, die Hohe Tatra und den Sudetennordrand im E. Die in diesem Bereich auftretenden Fluorit-Baryt-Lagerstätten sind überwiegend gangförmige Bildungen. Daneben treten noch Imprägnationen, metasomatische Verdrängungskörper und synsedimentäre Vorkommen auf. In tektonischer Hinsicht ist das abgegrenzte Gebiet, neben den Auswirkungen der alpidischen Faltung, durch das betonte Hervortreten der saxonischen Bruchtektonik gekennzeichnet.

Bedeutende Lagerstätten und bekannte Vorkommen finden sich vorwiegend dort, wo saxonische Brüche varizische Grundgebirgsschollen durchsetzen. Im einzelnen können folgende Lagerstättengebiete genannt werden (Abb. 1):

143

DDR: Harzscholle (Straßberg, Rottleberode, Kyffhäuser, Ilfeld, Mansfelder Rücken) (1); Flechtinger Höhenzug (2); Thüringer Wald (Schmalkalden, Ruhla, Ilmenau-Gehren, Kamsdorf-Könitz u. a.) (3); Vogtland (Schönbrunn, Brunndöbra u. a.) (4); Erzgebirge (Freiberg-Halsbrücke, Marienberg, Annaberg, Bärenstein u. a.) (5); des weiteren kleinere Vorkommen in den dazwischenliegenden Gebieten.

ČSSR: Erzgebirge (Teplice, Děčin, Moldava, Jachymov u. a.) (6); W-Sudeten (Harrachov, Křižany u. a.) (7); E-Sudeten (Gänge zwischen Olomouc und Opava) (8); Böhmerwald (Šumava) (9); Böhmisch-Mährisches Kristallin (10); Boskovicker Furche (11); Slowakei (12).

VR. Polen: Sudetennordrand (Wałbrzych, Stanislawow) (13).

BRD: W-Harz (Lauterberg, Andreasberg, Clausthal z. T. u. a.) (14); Brahmscher Massiv (15); Weserbergland-Egge-Geb. (16); Rheinische Masse (Ruhrbezirk, Brilon, Dreislar, Eifel) (17); Rheinpfalz (18); Richelsdorfer Geb. (19); Rhön (20); Spessart (21); Odenwald (22); Schwarzwald (23); Franken (24); Oberpfalz (Nabburg-Wölsendorf) (25); Bayrischer Wald (26).

Frankreich: Vogesen (Markirch) (27).

Österreich: Vorwiegend in der Grauwackenzone (Schwaz, Gailtal u. a.) und in der Mitteltrias. Letztere sind bevorzugt synsedimentär an die Wettersteinkalke gebunden und an deren Ausbiß am Alpennordrand über größere Bereiche bis in westdeutsches Gebiet zu verfolgen (28). Diese Vorkommen werden hier nicht mit in die nähere Betrachtung einbezogen

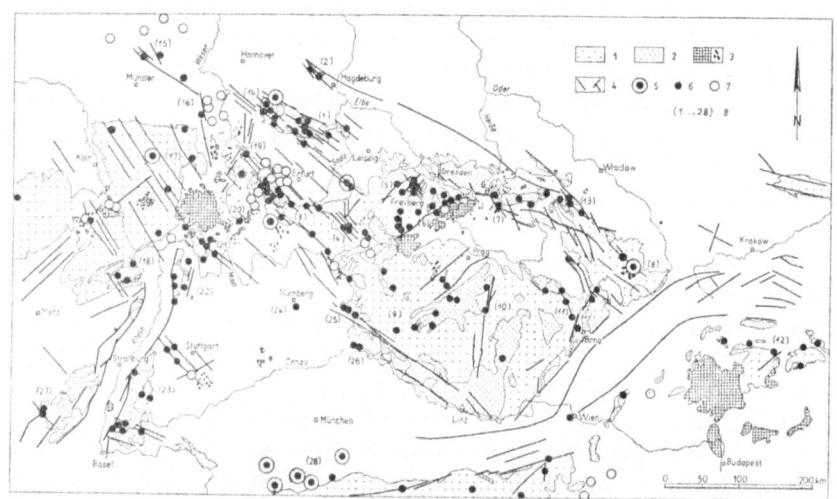

Abb. 1. Die Fluorit-Baryt-Lagerstätten Mitteleuropas.

1 — Paläozoikum; 2 — Variszische Magmatite; 3 — Postvariszische Effusiva; 4 — Störungen; 5 — Syngenetische F-Ba-Lagerstätten (stratiforme Lager und Imprägnationen); 6 — Epigenetische F-Ba-Lagerstätten (Gänge und metasomatische Bildungen); 7 — Tertiäre Kohlensäuerlinge; 8 — Lagerstättengebiete.

Der Gedanke, diese Lagerstätten unter einem einheitlichen Gesichtspunkt zu betrachten, ist nicht neu. So zeigte bereits *Mohs* (1931) gewisse regionale Zusammenhänge auf. Während *Schneiderhöhn* (1949, 1953) wieder mehr die Eigenständigkeit betonte, versuchte man in zahlreichen Arbeiten der letzten Jahre gemeinsame Züge zu finden und neu auszuarbeiten (*Werner* 1966; *Borchert* 1967; *Leeder* 1966, 1967; *Baumann* 1967, 1968; *Baumann* & *Rösler* 1967; *Baumann* & *Werner* 1968; *Baumann* & *Leeder* 1969; *Schröder* 1970, 1971 u. a.). Die neueren Untersuchungen stützen sich auf regional-geologische, tektonische, paragenetische, geochemische und geochronologische Daten, die zum größten Teil an den Lagerstätten der DDR gesammelt wurden.

2. Die Strukturformen der Lagerstätten

Wie aus der geologischen Situation der Lagerstättengebiete und ihrer räumlichen Verteilung hervorgeht (Abb. 1), sind die genannten Fluorit-Baryt-Lagerstätten bevorzugt an Kreuzungszonen von Tiefenstörungen bzw. an deren Kreuzungen mit den Faltungsstrukturen der paläozoischen Gebirgsrümpfe gebunden. Da diese Koppelung ein regionales Phänomen darstellt, das in ähnlicher Weise auch in den bedeutenden Lagerstättengebieten der Sowjetunion (Transbaikalien), der USA (Weststaaten) und Mexikos beobachtet werden kann, verdienen diese Zusammenhänge eine nähere Untersuchung.

2. 1 Geologisch-tektonische Stellung der Lagerstätten

Die großen geologischen Störungssysteme stellen Schwächezonen innerhalb der Erdkruste dar, die bei geeigneter Anlage des Kräfteplanes zu verschiedenen Zeiten aktiviert werden können. In Mitteleuropa existieren seit dem algonkischen Umbruch drei ausgeprägte Störungsrichtungen, die den tektonischen Bauplan bestimmen:

NE-SW (= erzgebirgische Richtung)
  Verlauf der variszischen Geosynklinal- und Antiklinalzonen; Verlauf der Faltenachsen, der Schieferung und bestimmter faltungstektonischer Kluftsysteme, Tiefenstörungen.
  Beispiele: Erzgebirgische Aniklinalzone, Spessart-Ruhla-Kyffhäuser-Zone, Zentraleuropäisches Lineament (mit Erzgebirgsabbruch), Boskovicer Furche u. a.
NW-SE (= herzynische Richtung)
  Hauptrichtung der saxonischen Störungen, die seit dem Zechstein und speziell in den kimmerischen Phasen im Rahmen der alpidischen Tektogenese angelegt bzw. erneut aktiviert wurden; weitgehend Schollen- und Bruchtektonik.
  Beispiele: Elbe-Lineament, Fläming-Linie, Harzer und Thüringer Randspalten, Bayerischer Pfahl u. a.

N-S (= rheinische und eggische Richtung)

Bruchelemente des Mittelmeer-Mjösen-Lineaments, Störungsaufspaltungen an Hochschollen, Klufthäufungszonen, geophysikalische Indikationen.

Beispiele: Rheintalgraben, Ohmgebirge, Plauen-Gera-Halle-Linie, Blanicka Furche u. a.

Bei der Verteilung der Fluorit-Baryt-Lagerstätten ist auffällig, daß an den Kreuzungsstellen dieser Störungslinien gewisse Häufungen auftreten. Das trifft in erster Linie für die NE-SW- und NW-SE-Systeme zu, läßt sich jedoch auch an den Kreuzungen dieser beiden Systeme mit den N-S-Strukturen erkennen.

Der Zusammenhang zwischen den Kreuzungen von Störungszonen und den Lagerstättenbildungen ist dreifältiger Natur:

a) Durch tiefgreifende Störungssysteme werden Magmenherde in tieferen Krustenbereichen oder im oberen Mantel angeschnitten und geben somit Anlaß zu Magmenbewegungen. Da die hydrothermale Lagerstättenbildung auf dem Vorhandensein und der Differentiation von intrudierenden Magmen beruht, ist damit eine erste genetische Bedingung erfüllt. Offensichtlich spielen für den Magmenaufstieg die Kreuzungszonen zwischen rhenotypen und herzynischen Strukturen eine besondere Rolle, da sich in Mitteleuropa die N-S-Richtungen durch auffällige geophysikalische Anomalien bemerkbar machen und darüberhinaus in diesen Kreuzungsbereichen während des Tertiärs simatische Magmen effusiv wurden.

b) Die oberflächennahen Störungssysteme, deren Bau weitaus komplizierter ist als der der Tiefenbrüche, geben durch Wechsel von Zerrung und Pressung sowie durch horizontale und vertikale Bewegungen Anlaß zur Bildung von Absatzräumen für aufsteigende Hydrothermen. Damit ist die Voraussetzung von raumschaffenden Prozessen für die Lagerstättengenese erfüllt. Dies gilt sowohl für Gang- als auch für metasomatische Lagerstättenbildungen. Selbstverständlich ist zwischen den raumschaffenden Prozessen für die Magmenintrusion und für die hydrothermale Lagerstättenbildung ein zeitlicher Hiatus eingeschaltet, der dem Magma die Möglichkeit zur Differentiation gibt.

c) Eine weitere wichtige Tatsache ist, daß von den möglichen Kreuzungspunkten besonders diejenigen lagerstättenführend sind, wo sich die Grundgebirgsschollen in einer Hochlage befinden. Daraus ergibt sich einerseits, daß ein Zusammenhang zwischen Magmenaufstieg und Hochschollenbildung besteht und damit die postmagmatischen Differentiate bevorzugt an oder in Horsten aufsteigen, andererseits sind die metamorphosierten und von palingenen Graniten durchsetzten kristallinen Grundgebirgsbereiche für eine tektonische Beanspruchung und Spaltenbildung wesentlich günstigere Gesteine als die weniger verfestigten mesozoischen Sedimente.

Die Wirkung dieser drei Einflußmomente läßt sich an den mitteleuropäischen Fluorit-Baryt-Lagerstätten deutlich erkennen. Allein ihre Existenz setzt eine magmatische Aktivität voraus, die sich im gegebenen regionalen Rahmen ausgewirkt haben muß. Die raumschaffenden Prozesse lassen sich in den Aufschlüssen der Ganglagerstätten belegen, wo teils Scherspalten mit weitgehend konstanter NW-SE-Richtung, teils auch deren Fiedersysteme mit flach- oder steilherzynem Streichen mineralisiert wurden. Die Kompliziertheit und zeitliche Gliederung der Bewegungen läßt sich an der Reihenfolge und der lokal wechselnden Intensität der einzelnen Abfolgen belegen. Ausnahmen bilden lediglich die Lagerstätten,

— die zum variszischen Geosynklinalmagmatismus gehören und submarin-hydrothermal in Sedimentationsräumen entstanden sind (z. B. Rammelsberg, Meggen, Ostsudeten);
— die postorogen im Zusammenhang mit variszischen palingenen Graniten entstanden sind (z. B. pneumatolytische Fluoritmineralisationen auf Zinnlagerstätten, Teile der hydrothermalen Ganglagerstätten des Harzes, des Erzgebirges, der Böhmischen und Rheinischen Masse u. a.);
— die metasomatisch in Kalken gebildet wurden, obwohl auch hierzu Spalten für die Substanzzufuhr und die Gewährleistung der Wegsamkeit vorhanden gewesen sein mußten.

Die bevorzugte Bindung der Lagerstätten an Gesteine des Grundgebirges läßt sich an vielen Stellen belegen. Entweder sind Gebiete mit Kreuzungsstellen im Bereich des mesozoischen Deckgebirges lagerstättenfrei, wie es im Falle des Thüringer Beckens, des Subherzyns oder im Bereich der schwäbisch-fränkischen Trias zu beobachten ist, oder die Lagerstätten verarmen substantiell und zahlenmäßig sobald die Störungen aus dem Grundgebirge in das jüngere Deckgebirge übergehen. Beispiele dafür finden sich an den Harzrändern, in der westlichen Verlängerung des Thüringer Waldes und speziell im östlichen Spessartvorland.

2. 2 Zeitliche Position der Lagerstättenstrukturen

Die zeitliche Einstufung der an der Lagerstättenbildung beteiligten Störungen kann prinzipiell auf 3 Arten erfolgen:

— nach der geologischen Datierung anhand von Diskordanzen oder Versetzung von altersbekannten Schichten bzw. die kausale Zuordnung nicht datierbarer Störungen zu den erstgenannten;
— nach der physikalischen Altersbestimmung an Mineralneubildungen im Störungsbereich; dazu bieten sich speziell K-Minerale (z. B. Serizite) zur Bestimmung nach der Kalium-Argon-Methode an;
— nach absoluten Altersdatierungen an den Gangmineralen selbst, unter der berechtigten Voraussetzung, daß zwischen Spaltenbildung und Spaltenausfüllung kein längerer Zeitraum vorhanden ist. Zur Bestimmung lassen sich die Pb-Pb-Methode, die Schwefelisotopenmethode, die U/Pb-Methode, Emanationsmessungen und die Methode des remanenten Magnetismus an Fe-Oxiden einsetzen.

Alle genannten Methoden wurden an verschiedenen Fluorit-Baryt-Lagerstätten angewendet und ergaben mit weitgehender Übereinstimmung eine Häufung der Alterswerte im Bereich von 150—180 Ma. *(Baumann & Rösler, 1967)*. Damit fällt der Zeitraum der Lagerstättenbildung im wesentlichen in die Wende Trias-Jura (kimmerische Phasen). Einzelne Werte streuen bis in den Bereich der subherzynen Phase (100 MA) einerseits und andererseits in den Bereich Zechstein-Trias. Eine weitere Häufung von Alterswerten in einigen Lagerstättengebieten zeigt tertiäres Alter an (Badenweiler, Jilové-Roztoky).

2. 3 Tektonische Lagerstättentypen

Die F-Ba-Lagerstätten und -Vorkommen Mitteleuropas treten in verschiedenen tektonischen Typen auf, die man sowohl nach dem Alter als auch nach der Struktur gliedern kann:

Lager

Zu den variszischen, syngenetischen Lagern gehören die Polymetall-Baryt-Lagerstätten des Rammelsberges bei Goslar und von Meggen im Sauerland. Weitere Vorkommen finden sich in den Ostsudeten (Jesenik-Geb.) (Abb. 1). Die Stoffzufuhr erfolgte auf Spalten aus Herden des geosynklinalen Magmatismus, der Absatz der Erze dagegen synsedimentär im Mitteldevon. Die Bildungen wurden anschließend in die variszische Faltung einbezogen und mehr oder weniger intensiv metamorphosiert (z. T. mit epigenetischen Umlagerungs- und Mobilisationsbildungen).

Saxonische syngenetische Vorkommen sind an den Plattendolomit des Zechsteins (Caaschwitz bei Gera) und an die Wettersteinkalke der mittleren alpinen Trias gebunden (Abb. 1). Die letztgenannten Bildungen haben im Bereich der Nördlichen Kalkalpen (Innsbruck, Nassereith, Lech, Füssen, Bad Tölz, Kufstein) Ausmaße von Lagerstätten. Auch hier erfolgte die Zufuhr vermutlich auf Spalten, der Absatz synsedimentär.

Metasomatische Bildungen

Metasomatische Fluorit-Baryt-Vorkommen variszischen Alters sind relativ selten und spielen wirtschaftlich keine Rolle. Es handelt sich entweder um pneumatolytisch-katathermale Greisenbildungen in Feldspatgesteinen (z. B. Ehrenfriedersdorf/Erzgebirge) oder um Kalkverdrängungslagerstätten (z. B. Breitenbrunn im Westerzgebirge).

Saxonische metasomatische Lagerstätten sind an Riffkalke des Zechstein (Leutnitz/Thüringen) und an Steinmergelbänke des mittleren (Gips-)Keupers (Römhild) gebunden.

148

Bei den reinen Ganglagerstätten sind sowohl im variszischen als auch im saxonischen Zyklus folgende Gangtypen zu unterscheiden:

— Scherspalten. Diese dienen für die Hydrothermallösungen bevorzugt als Aufstiegswege aus den tiefliegenden Herden. Sie sind tiefgreifend mineralisiert und zeigen selbst bei Vorherrschen von linsigen Erzmitteln langes Aushalten im Streichen. Dazu gehören ein großer Teil der Randspalten der Grundgebirgsschollen.
— Zug- und Fiederspalten. Diese wirken als tektonische Elemente niedrigerer Ordnung stärker lösungsverteilend. Sie können eine größere Mächtigkeit und auch eine intensivere Mineralisation erreichen als die zugehörigen Scherspalten.
— Zusammengesetzte Gänge. Werden Scherspalten mehrmals intensiv bewegt, so können sie als mächtige Ruscheln oder Breccienzonen einen eigenen Typ von Ganglagerstätten bilden. Je nach Lage der Öffnungsvektoren können parallel- bzw. diagonalstreichende Trümer entstehen, die oftmals verschiedene Paragenesen enthalten. Ein schönes Beispiel dafür ist der Straßberg-Neudorfer Gangzug des Unterharzes.

3. Die Mineralisation der Fluorit-Baryt-Lagerstätten

3. 1 Mineralparagenesen

Ein wichtiges Argument für die Konzeption einer einheitlichen Genese des größten Teiles der mitteleuropäischen Fluorit-Baryt-Lagerstätten ist durch ihre auffällig gleichartigen Paragenesen gegeben. Es handelt sich um weitgehend mineralische Fluorit- bzw. Barytgänge oder fluorit-barytbetonte Gänge mit wechselnden Anteilen an Karbonaten (Siderit, Ankerit, Dolomit und Calcit), Sulfiden (Pyrit, Chalkopyrit, Sphalerit, Galenit u. a.), Oxiden (Hämatit, Mn-Oxide) und Quarz.

Durch die bisherige uneinheitliche Bearbeitung war ein Vergleich nicht immer möglich, zumal aus verschiedenen Gründen bestimmte Minerale besonders hervorgehoben oder auch vernachlässigt worden sind.

Auf Grund eines regionalen Vergleichs von rund 150 Fluorit-Baryt-Lagerstätten Mitteleuropas wurde versucht, die Mineralparagenesen auf einen allgemeingültigen Standard zu beziehen *(Baumann & Leder, 1969)*. Für die saxonischen Fluorit-Baryt-Lagerstätten eigneten sich als Basis dazu besonders die Paragenesenschemata des Thüringer Waldes, des Erzgebirges und des Harzes, die einerseits in sich weitgehend übereinstimmen und andererseits bestimmte lokale Besonderheiten deutlich erkennen lassen *(Baumann & Werner, 1968)*. Trotz der lokalen paragenetischen, strukturellen und geochemischen Besonderheiten ergibt sich ein deutlicher minerogenetischer Zusammenhang aller saxonischen Fluorit-Baryt-Lagerstätten Mitteleuropas, der sich insbesondere in einer weitgehenden Gemeinsamkeit der Elementkombination sowie der Mineralvergesellschaftungen und -sukzessionen widerspiegelt. Im wesentlichen lassen sich die saxonischen Mineralisationen in folgende Hauptphasen untergliedern (Abb. 2):

Quarzige Eisen-Baryt-Phase: Quarz, Baryt, Fe-Mn-Oxide, Karbonate (= eba-Typ)*;

Fluorit-Baryt-Phase: Baryt, Fluorit, Quarz, z. T. Fe-Cu-Pb-Zn-Sulfide (= fba-Typ);

BiCoNi-Phase: Quarz, Baryt, Fluorit, Co-Ni-Fe-Arsenide, z. T. Ag-, Bi- und U-Minerale (= BiCoNi-Typ);

Karbonatische Ag-S-Phase: Karbonspäte, Quarz, Fe-Cu-Pb-Zn-Sulfide (z. T.), Ag-Hg-Sb-As-Minerale (= Edle Geschicke-Typ);

Eisen-Mangan-Phase: Quarz, Fe-Mn-Oxide (= Fe-Mn-Typ).

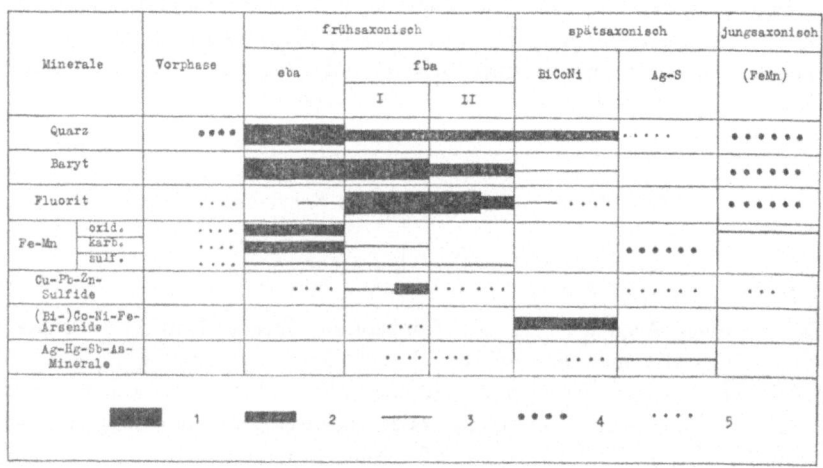

Abb. 2. Allgemeines Paragenesenschema (Standard) der saxonischen Fluorit-Baryt-Lagerstätten (aus *Baumann* & *Leeder*, 1969)
Intensität-Extensität: 1 — Sehr stark; 2 — Stark; 3 — Mittel; 4 — Schwach; 5 — Sporadisch.

Diese fünf Mineralisationsphasen lassen sich in fast allen saxonischen Lagerstättenbezirken Mitteleuropas nachweisen, wobei die Intensität der einzelnen Phasen (= Gangformationen) regional unterschiedlich ist. So sind die einzelnen Formationstypen in folgenden Bezirken besonders betont:

* Die in Klammer angeführten Typenbegriffe entsprechen den klassischen Gangformationsbezeichnungen des Erzgebirges.

eba-Typ: Erzgebirge (DDR- und ČSSR-Seite), Ostthüringen, Harz (Ilfelder Becken), W-Sudeten, Oberpfalz (Nabburg), Franken, Spessart-Odenwald, Schwarzwald, Slowakei;

fba-Typ: — mit wenig Sulfiden
Flechtingen, E-Harz, Thüringer Wald, Richelsdorfer Geb., Erzgebirge (ČSSR-Seite), N-Sudeten, W-Sudeten, Böhmisch-Mährisches Kristallin, Oberpfalz, Franken, Bayerischer Wald, Spessart-Odenwald, Rheinpfalz, Schwarzwald;
— mit viel Sulfiden
E-Erzgebirge (Freiberg-Halsbrücke), N-Harz, W-Harz, Sudeten, Mittelböhmen, Rheinisches Schiefergebirge, Schwarzwald (N-Teil), Alpennordrand (z. T.);

BiCoNi-Typ: W-Erzgebirge (DDR- und ČSSR-Seite), W-Sudeten, Thüringer Becken (Mansfelder Rücken), Richelsdorfer Geb., W-Harz, Rheinisches Schiefergebirge, Schwarzwald (N-Teil), Oberpfalz;

Ag-S-Typ: Erzgebirge, W-Harz, Böhmisch-Mährisches Kristallin, Oberpfalz, Richelsdorfer Geb., Rheinpfalz;

Fe-Mn-Typ: Erzgebirge (DDR- und ČSSR-Seite), Harz, Thüringer Wald, Böhmisch-Mährisches Kristallin, Oberpfalz, Bayerischer Wald, Schwarzwald (Rheintalgraben), Brahmscher Massiv.

Aus der Vorherrschaft einzelner oder aller Formationstypen kann man eine gewisse regionale Zonalität ableiten, die auf gewisse Zentren der Mineralisation schließen läßt. Je tiefer bzw. größer der Grundgebirgsanschnitt ist, umso deutlicher tritt die Gesamtheit aller Formationstypen in Erscheinung (Erzgebirge, Thüringen, Harz, Böhmisch-Mährisches Kristallin, Sudeten, Schwarzwald). Diese Bereiche stellen gleichzeitig die Hauptzentren der saxonischen Mineralisation dar. Demgegnüber sind im Bereich der kleinen Horstbildungen bzw. in den Zwischengebieten der Grundgebirgsschollen oftmals nur einzelne Formationstypen ausgebildet (Flechtinger Scholle, Richelsdorfer Geb., Spessart, Odenwald, N-Sudeten). Des weiteren ist beim regionalen Vergleich der Mineralparagenesen der jeweilige Charakter des Nebengesteins (strukturell und stofflich) zu berücksichtigen. So können in kalkigen Gesteinen metasomatische Verdrängungen auftreten (z. B. Schmalkalden und Leutnitz/Thür.). Entsprechend dem herrschenden Redoxpotential im Ausscheidungsbereich sind Veränderungen im Mineralcharakter zu beobachten. So kann z. B. das Fe als Hämatit (oxydisch), als Siderit (karbonatisch) oder als Pyrit-Markasit (sulfidisch) abgeschieden werden. Dabei sind die oxydischen Bildungen bevorzugt an saure Effusiva (Ilmenau, Ilfeld, N-Sudeten) oder kristalline Schiefer (Erzgebirge) und die karbonatischen Bildungen an pelitisch-psammitische Sedimentgesteine gebunden (Ostthüringen, Harz). Allgemein bekannt ist auch die topomineralische Einflußnahme auf die Ausscheidung der BiCoNiAg-Formation (durch bitumiöses Nebengestein: Mansfeld, Richelsdorf; durch basische Metamorphite: Erzgebirge).

3. 2 Mineralisationsepochen

Hinsichtlich der absoluten zeitlichen Eingliederung der Fluorit-Baryt-Mineralisationen wurden nach den älteren Ansichten, insbesondere nach *Schneiderhöhn* (1941), alle Gänge, unabhängig von ihrer Paragenese und ihrer strukturellen Position, zum postmagmatischen Ganggefolge der variszischen Granite gerechnet. In einigen Fällen, wie im Schwarzwald, im Erzgebirge und im Vogtland, ließen sich scheinbar einwandfreie Zonalitäten zu Granitoberflächen zeigen. Ein Musterbeispiel dafür war die Lagerstättenzonalität um den Ramberggranit des Harzes.

Andererseits wurden alle Lagerstätten, die in jüngeren Schichten auftreten oder abweichende Paragenesen zeigen und einer derartigen Einordnung nicht entsprechen, für sekundär-hydrothermale Bildungen angesehen *(Schneiderhöhn* 1949, 1953). Diese Vorstellung war in vielen Fällen unlogisch und hielt einer eingehenderen Überprüfung nicht stand. Auf Grund umfangreicher Untersuchungen der letzten Jahre ergibt sich, daß in den meisten Lagerstättenbezirken Mitteleuropas zwei verschiedenartige und genetisch unterschiedliche Mineralisationszyklen (variszisch und postvariszisch = saxonisch) unterschieden werden können. Eine wesentliche Ergänzung der paragenetischen Untersuchungen waren dabei vor allem absolute Altersdatierungen (U-Pb, K-Ar, Pb-Pb), isotopengeochemische ($^{32}S/^{34}S$, $^{18}O/^{16}O$) und geochemische Bestimmungen *(Baumann & Rösler* 1967, *Baumann & Leeder* 1969 u. a.) sowie systematische Untersuchungen an Flüssigkeitseinschlüssen in Fluoriten *(Baumann, Harzer & Leeder,* 1972).

Die Ergebnisse lassen keinen Zweifel an der absoluten zeitlichen Einstufung der Fluorit-Baryt-Lagerstätten Mitteleuropas. Außer den synsedimentären devonischen Geosynklinallagerstätten vom Typ Rammelsberg-Meggen (Ba) und den mit geringer Intensität auftretenden Mineralisationen in der pneumatolytischen (F) und hydrothermalen Ganggefolgschaft (Ba) der variszischen Granite sind die bedeutenden F-Ba-Mineralisationen in den saxonischen (alpidischen) minerogenetischen Zyklus einzuordnen. Dabei kann die saxonische Mineralisation bereits im Zechstein einsetzen, wie die synsedimentären Fluorite von Caaschwitz bei Cera vermuten lassen. Die Hauptbildungszeit fällt jedoch in die kimmerische und subherzyne Phase des Mesozoikums. In dieser Zeit bildeten sich sowohl die saxonischen Ganglagerstätten als auch die syngenetischen Vorkommen im Wettersteinkalk der nördlichen und südlichen Kalkalpen und im Keuper von Hildburghausen sowie die metasomatischen Vorkommen im Zechstein von Thüringen.

Die im Tertiär gebildeten Gangvorkommen treten in ihrer Bedeutung weit zurück. Sie sind jedoch genetisch sehr interessant, zumal in der Umgebung von Teplice auch eine rezente und subrezente Fluoritabscheidung zu beobachten ist. Die tertiären bis rezenten Kohlensäuerlinge sind als letzte Aktivitäten dieses Mineralisationszyklus anzusehen (Abb. 1).

4. Zur Genese der Fluorit-Baryt-Lagerstätten

4. 1 Zur Herkunft der Lösungen

Die Anerkennung einer weitgehenden strukturellen und paragenetisch-geochemischen Übereinstimmung der mitteleuropäischen Fluorit-Baryt-Lagerstätten setzt einen gemeinsamen und stofflich identischen Ursprungsherd voraus. Bei der regional sehr ausgedehnten Mineralprovinz von der Rheinischen Masse bis an die mährisch-slowakische Grenze und von der Flechtinger Scholle bis an den Alpennordrand kann es sich natürlich nicht um einen einzigen Ursprungsherd handeln, sondern nur um mehrere Herde, die im gleichen Zeitraum unter gleichen Bedingungen aktiv gewesen sind. Die postmagmatische Tätigkeit palingener Granite scheidet als Quelle aus, da Granite weder zeitlich, räumlich noch stofflich in die saxonische Tektogenese eingeordnet werden können. Eine Spätwirkung der variszischen Granite ist aus kalorischen und petrogenetischen Gründen ebenfalls sehr unwahrscheinlich.

Seit Beginn der eingehenderen Untersuchung von Fluorit-Baryt-Lagerstätten, speziell im Raum von Thüringen, wurden eine Reihe von Bildungshypothesen entwickelt (*Bärtling* 1911; *Staub* 1928, 1929; v. *Engelhardt* 1936; *Thienhaus* 1941; *Gimm* 1947; *Oelsner* 1956 u. a.), die jedoch aus tektonischen, geochemischen, paragenetischen, lagerstättenkundlichen oder zum Teil auch aus rein logischen Gründen verworfen werden mußten (*Werner* 1958, 1966; *Baumann* & *Rösler* 1967; *Leeder* 1967 u. a.).

Ausgehend von den geotektonischen Vorstellungen über die Wechselwirkung zwischen Erdmantel und -kruste kann festgestellt werden, daß Bildung und Umbildung der Sialschale kausal durch säkular-plastische Massenverlagerungen im Mantel bedingt sind. Bruchreaktionen der spröden Kruste geben Anlaß zu Druckentlastungen, Magmenbildung und Magmenaufstieg aus dem Mantelbereich. Bevorzugte Gelegenheiten für den Aufstieg simatisch-juveniler Magmen sind einmal die Geosynklinaleinsenkungen mit ihren Randbrüchen und zum anderen die quasikratone Bruchtektonik in den Tafelbereichen. Damit besteht im Anfangs- und Endstadium (Konsolidierungsstadium) eines orogenetischen Zyklus die Möglichkeit zu Intrusionen und Effusionen juveniler Magmen. Auch die autonomen Tafel-Aktivierungen sind kein steriler Prozeß, wie bisher häufig angenommen wurde, sondern können mit intensiven magmatisch-lagerstättenbildenden Vorgängen verbunden sein (*Ščeglov* 1969). So sind die Ganglagerstätten Transbajkaliens, des Aldan, Bulgariens, der westlichen USA-Staaten und vieler anderer Gebiete im wesentlichen an ausgeprägte Bruchzonen gebunden, wie überhaupt 85 % aller Fluoritlagerstätten der Welt von Tiefenbrüchen kontrolliert werden (*Gruškin* 1964).

Im Falle von Intrusionen neigen simatische Schmelzen zur weitgehenden gravitativen Differentiation, die unter Ausscheidung schweren Materials in den tieferen Bereichen (ultrabasischer und basischer Gesteine) zu alkalireichen, atlantischen

Magmen in den oberen Zonen führt. Die dabei auftretende Anreicherung von leichtflüchtigen Stoffen, leichten Elementen und speziell von Alkalien fördert die weitere Differentiation, die unter günstigen Umständen bis zu extremen Fraktionen führen kann. Für derartige reife Differentiationskomplexe ist die Kombination von Alkaligesteinen, wie Alkali- und Nephelinsyenite, Ijolithe, Melteigite, Foyaite und vor allem Karbonatite, neben Oliviniten, Pyroxeniten und Alkaliultrabasiten typisch. Es ist weiterhin wichtig, daß speziell in den leichtesten Alkaligesteinsdifferentiaten vor allem die Elemente angereichert werden, die in den Paragenesen der saxonischen Lagerstättenbildungen auftreten: Ca-Mg-Ba-Sr, Fe-Mn, Pb-Zn-Cu, F, SO_4, S, CO_2 und als Spurenelemente Bi, Co, Ni, SE, U, Hg, Sb und As. Aus vielen Karbonatitvorkommen ist bekannt, daß neben den Karbonaten vor allem Fluorit und Baryt bis zu Lagerstättendimensionen angereichert sein können. Bisher wurden im Weltmaßstab rund 330 Karbonatitkomplexe, besonders in aktivierten Tafelbereichen, nachgewiesen. Mit diesen simatogenen Differentiationsbildungen sind bedeutende Lagerstätten von Calcit-Dolomit, Fluorit, Baryt, Apatit, Magnetit sowie zahlreiche anderer Minerale, u. a. auch Fe-Ni-As-S-Minerale sowie Cu-Pb-Zn-Sulfide, verbunden *(Smirnov* 1970; *Tuttle* & *Gittins* 1965 u. a.). Auch die Fe-Mn-Ba- und Pb-Zn-Cu-Ba-Bildungen des jungen aktiven Riftsystems des Roten Meeres werden mit einem simatischen Magmatismus in Verbindung gebracht *(Bonatti* u. a., 1972).

Überträgt man diese weltweit zu beobachtenden Tatsachen auf die lokalen Verhältnisse Mitteleuropas, so ergeben sich folgende Schlußfolgerungen:

Die Alpengeosynklinale und die nördlich angrenzenden variszisch konsolidierten Tafelgebiete Mitteleuropas unterliegen während des Mesozoikums-Känozoikums weitgehend einer gemeinsamen tektonischen Beanspruchung, die im mitteleuropäischen Raum vorwiegend germanotyp (= saxonische Tektogenese) in Erscheinung tritt. Die durch die tiefreichenden Bruchzonen aktivierten simatischjuvenilen Magmen vermögen sich zu differenzieren und im Verlauf der gravitativen Stofftrennung lagerstättenbildende Lösungen bereitzustellen. Derartige Differentiationsherde werden sich dort bilden, wo die tektonischen Bedingungen zum Aufstieg günstig sind. Das wird auf bedeutenden Störungen mit entsprechendem Tiefgang oder auf Kreuzungsstellen derartiger Strukturen der Fall sein. Erneute Bewegungen auf diesen Störungen geben dann Anlaß zum Aufstieg postmagmatischer Lösungen. Die Messung der Homogenisierungstemperaturen an Flüssigkeitseinschlüssen von Fluoriten des Harzes und des Erzgebirges führten zu Bildungsbereichen von 400—30⁰ C bei altersmäßiger Abstufung der einzelnen Fluoritgenerationen. Das Maximum der gemessenen Temperaturen liegt dabei zwischen 90⁰ und 160⁰ C (Abb. 3). Diese Messungen wurden durch Sauerstoffisotopen-Untersuchungen an Karbonaten und an Quarzen der gleichen Paragenesen ergänzt. Dabei ergaben sich z. B. bei den Temperaturmessungen über die Isotopenverhältnisse an den Karbonaten nur vernünftige Werte, wenn sie auf Wässer von magmatisch-juveniler Herkunft modelliert wurden. Die Modellierung der Isotopenzusammensetzung sowohl der Karbonate als auch der Quarze auf rein magmatisches Wasser führte zu Bildungstemperaturwerten, die mit den aus den Homogenisierungsmessungen gewonnenen Temperaturen vollkommen übereinstimmten. Aus den gemessenen ^{18}O-Werten ließ sich somit der Schluß ziehen, daß hinsichtlich des Ursprungs der mineralbildenden Hydrothermen der postvariszischen (saxonischen) Fluorit-Baryt-Lagerstätten ein primärmagmatischer, d. h. simatisch-juveniler Herd angenommen werden kann *(Baumann, Harzer u. Leeder 1972).*

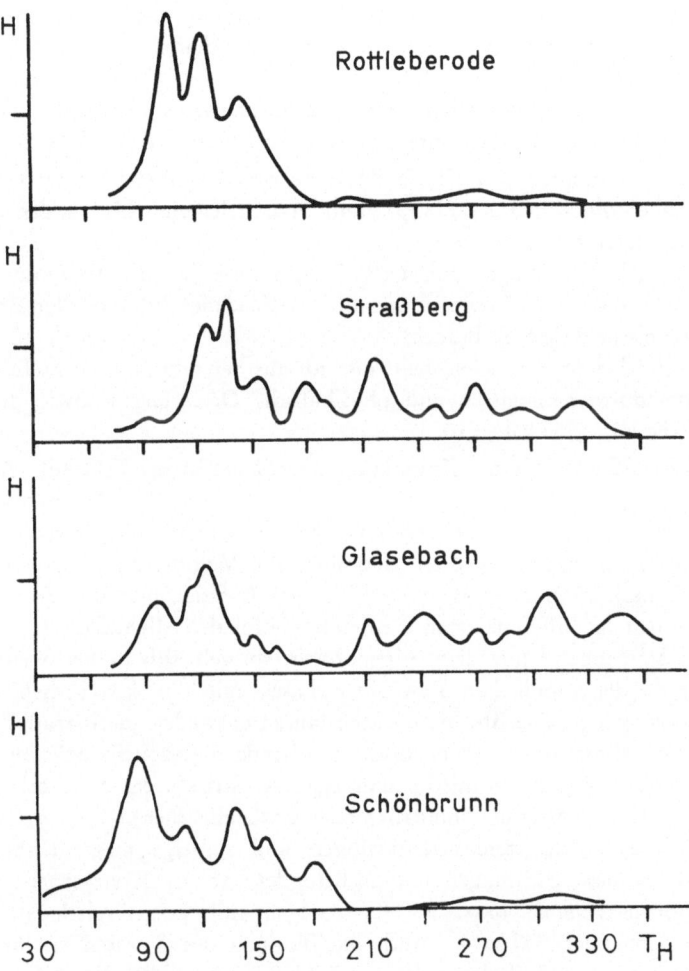

Abb. 3. Häufigkeitsdiagramm der Homogenisierungstemperaturen (T_H) an Flüssigkeitseinschlüssen von Fluoriten einiger Lagerstätten des Harzes und des Erzgebirges (aus *Baumann, Harzer* und *Leeder,* 1972).

4. 2 Die minerogenetische (metallogenetische) Stellung der Fluorit-Baryt-Lagerstätten in Mitteleuropa

Die Fluorit-Baryt-Lagerstätten und -Vorkommen Mitteleuropas lassen sich zwei minerogenetischen Zyklen zuordnen: dem geosynklinal-orogenen variszischen und dem kratogen-bruchtektonisch betonten postvariszischen (saxonischen) Zyklus. Innerhalb beider Zyklen treten syngenetisch-sedimentäre Vorkommen hydrothermaler Herkunft, hydrothermale Gangfüllungen und aszendent-metasomatische Bildungen auf. Die weitaus größte Zahl der Vorkommen gehört zum saxonischen Zyklus i. e. S., der die tektonischen Hauptphasen des Mesozoikums umfaßt. In diesem Zyklus werden auch die wirtschaftlich wichtigsten Fluorit-Baryt-Lagerstätten gebildet.

Diese saxonischen Lagerstätten haben drei gemeinsame Merkmale, die die Zuordnung zu einer einheitlichen minerogenetischen Provinz rechtfertigen:

a) eine übereinstimmende stoffliche Zusammensetzung, die sich von den Paragenesen über die Minerale, Haupt- und Spurenelemente bis zu den Isotopenwerten erstreckt;

b) eine weitgehend analoge strukturelle Position, die sie trotz der Unterschiede in der Form und im Auftreten innerhalb verschiedener lithofazieller Bereiche an die saxonische Tektonik bindet;

c) eine weitgehende Altersgleichheit, die sich aus der tektonischen Stellung ergibt und die durch geologische und physikalische Datierungen sowie durch geochemische Kriterien belegt ist.

Der wesentliche Faktor der Kausalzusammenhänge ist die Tektonik. Ausgehend von den geotektonischen Vorgängen, die zur Anlage der Alpengeosynklinale führten, reicht der Einfluß der Tektonik über die Anlage von Störungssystemen, z. T. in persistenten Lineamenten, und über die Magmenintrusionen auf diesen bis zu den raumschaffenden Prozessen für den Aufstieg und die Verteilung der Hydrothermen aus differenzierten Herden sowie für den Mineralabsatz.

Der Zusammenhang mit dem Alpenorogen spiegelt sich in der weitgehenden Zeitgleichheit der alpidischen Bewegungsphasen mit der saxonischen Tektonik Mitteleuropas wider. Die Bruchstrukturen laufen entweder quasiparallel zur alpidischen Geosynklinalfurche (herzynisch streichende mesozoische Störungssysteme) oder normal zu dieser (rheinische und eggische Strukturen, z. T. erkennbar in Form von geophysikalischen Anomalien oder Grabenbrüchen).

Ausgeprägte Lagerstättenkonzentrationen und sonstige magmatische Aktivitäten sind an diese Richtungen und insbesondere an die Kreuzungsbereiche der beiden Störungssysteme bzw. an Kreuzungen mit erzgebirgisch streichenden Strukturen gebunden (Abb. 4). Auch die Bindung des jüngsten tertiären simatischen Magmatismus und der rezenten Kohlensäuerlinge an diese Kreuzungsbereiche ist evident.

Die weiterhin offensichtliche Bindung der Lagerstätten an Grundgebirgsschollen darf nicht dahingehend ausgelegt werden, daß eine genetische Beziehung

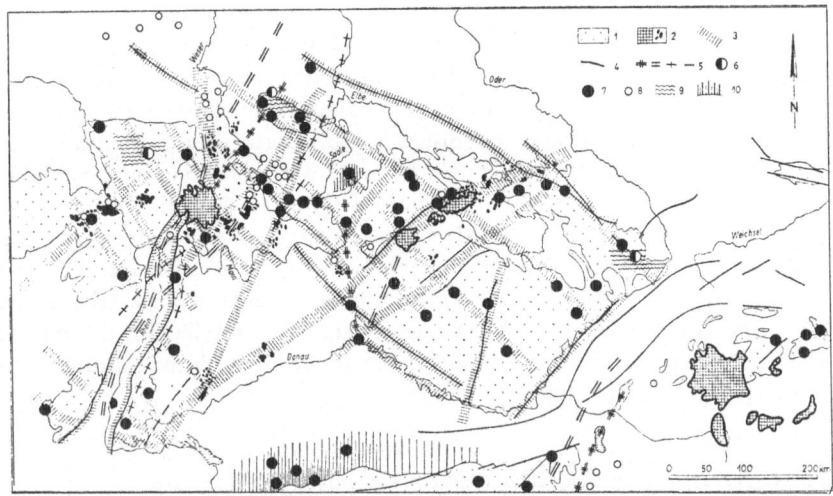

Abb. 4. Beziehungen zwischen tektonischen Strukturen, postvariszischem Magmatismus und Verbreitung der Fluorit-Baryt-Lagerstätten in Mitteleuropa.
1 — Paläozoisches Grundgebirge; 2 — Postvariszische Effusiva; 3 — Störungszonen; 4 — wichtige Tiefenstörungen; 5 — Geophysikalische Anomalien; 6 — Variszische F-Ba-Lagerstätten; 7 — Saxonische F-Ba-Lagerstätten (z. T. mit variszischen Mineralisationen); 8 — Kohlensäuerlinge (Tertiär); 9 — Paläozoische Schichten mit syngenetischer Mineralisation; 10 — Mesozoische Schichten mit syngenetischer Mineralisation.

zwischen paläozoischen Gesteinen oder Magmatiten und diesen Lagerstätten vorhanden ist. Der Zusammenhang besteht lediglich darin, daß im Zuge der tektonischen Zerstückelung des Tafelgebietes, der dadurch bedingten Vertikalbewegungen der Teilschollen und der magmatischen Prozesse im Untergrund Hoch- und Tiefschollen entstehen, wobei erstere die heutigen Grundgebirgsanschnitte darstellen. Die Beziehung Lagerstätte-Grundgebirge ist nur mechanisch bedingt, indem die kristallinen Gesteine dieser Bereiche günstige Möglichkeiten zur Spalten- und Raumbildung bieten. Eine Bestätigung dafür ist in der Vertaubung der lagerstättenführenden Strukturen beim Verlassen der Schwellengebiete und in der Mineralisationsfreiheit der gleichen Störungen in den Beckenbereichen (Tiefschollen) zu sehen. Daraus ergibt sich, daß alle herzynischen Störungen im Bereich von Grundgebirgserhebungen generell als lagerstättenhöffig anzusehen sind.

Die teritären Mineralvorkommen gleicher Paragenese stellen ein jüngeres Äquivalent zu den saxonischen Lagerstätten dar. Sie stehen zeitlich mit der Endphase der alpidischen Orogenese in Verbindung, die sich kinetisch auch im konsolidierten Vorland auswirken mußte. Mit diesen Bewegungen sind analoge, wenn auch im Umfang bescheidenere magmatische und postmagmatische Prozesse verbunden. Daher sind auch die tertiären Störungen im Berich des begleitenden Magmatismus als lagerstättenhöffig anzusehen.

Neben diesen strukturell kontrollierten intrakrustalen Fluorit-Baryt-Lager-

157

stätten besteht noch die Möglichkeit eines submarinen Hydrothermenaustritts und einer syngenetischen Ausscheidung der Mineralsubstanz im Rahmen der marinen Sedimentation. Es kommt dann zu den submarin-hydrothermalen und (sekundär)-metasomatischen Vorkommen in verschiedenen sedimentären Schichten des Mesozoikums-Känozoikums (Zechstein: Caaschwitz/Thür.; Buntsandstein: Brahmscher Massiv; Wettersteinkalk: Vorkommen der Nördlichen und Südlichen Kalkalpen, Gorny Šlask; Keuper: „Bleiglanzbänke" von Römhild/Thür., Hildburghausen; Kreide: Dollbergen/Niedersachsen). Auf Grund der sehr ähnlichen Elementkombinationen und Mineralparagenesen (F, ± Ba, Fe, Pb, Zn, Cu) ist es sehr wahrscheinlich, daß diese mesozoischen hydrothermal-sedimentären Lagerstätten einerseits und die saxonischen Ganglagerstätten (einschließlich der metasomatischen) andererseits genetische Äquivalente in unterschiedlichen strukturellen und lithologischen Verhältnissen darstellen.

Literatur

Bärtling, R. (1911): Die Schwerspatlagerstätten Deutschlands. — Stuttgart.

Baumann, L. (1967): Zur Frage der varistischen und postvaristischen Mineralisation im sächsischen Erzgebirge. — Freib. Forsch.-H, C 209, 15—36.

Baumann, L. (1968): Die Mineralparagenesen des Erzgebirges — Charakteristik und Genese. — Freib. Forsch.-H. C 230, 217—233.

Baumann, L. & H. J. *Rösler* (1967): Zur genetischen Einstufung varistischer und postvaristischer Mineralisationen in Mitteleuropa. — Bergakademie, 16, H. 11, 660—664.

Baumann, L. & C.-D. *Werner* (1968): Die Gangmineralisation des Harzes und ihre Analogien zum Erzgebirge und zu Thüringen. — Ber. deutsch. Ges. geol. Wiss., Reihe B, 13, H. 5, 525—548, Berlin.

Baumann, L. & O. *Leeder* (1969): Paragenetische Zusammenhänge der mitteleuropäischen Fluorit-Baryt-Lagerstätten. — Freib. Forsch.-H. C 266 (Top. Rep. of IAGOD, Vol. I), 89—99.

Baumann, L., D. *Harzer* & O. *Leeder* (1972): Beitrag zum Charakter mineralbildender Lösungen in einigen Hydrothermallagerstätten der DDR. — Ber. deutsch. Ges. geol. Wiss., Reihe B, 17, H. 3, 341—355, Berlin.

Bonatti, E., D. E. *Fisher*, O. *Joensum*, H. S. *Rydell* & M. *Beyth* (1972): Iron-Manganese-Barium deposits from the northern Afar-Rift (Ethiopia). — Econ. Geol. 67, 717—730.

Borchert, H. (1967); Genetische Unterschiede zwischen varistischen und saxonischen Lagerstätten Westdeutschlands und deren Ursachen. — Freib. Forsch.-H. C 209, 47—63.

Chrt, J., H. *Bolduan* & al. (1966): Die postmagmatische Mineralisation des Westteils der Böhmischen Masse. —Sbornik geol. věd. ř. LG. 8, 113—192.

Chrt, J. & H. *Bolduan* (1967): Die postmagmatische Mineralisation des Westteils der Böhmischen Masse. — Freib. Forsch.-H. C 209, 39—46.

Chrt, J. (1968): Fluorite Deposits in Czechoslovakia. — Geologický průzkum, Vol. X, H. 7—8, 66—67.

Chrt, J., H. *Bolduan*, K.-H. *Bernstein*, J. *Legierski* & al. (1968): Räumliche und zeitliche Beziehungen der postmagmatischen Mineralisation der Böhmischen Masse zu Magmatismus und Bruchtektonik. — Z. angew. Geol., 14, H. 7, 362—376.

Engelhardt, W. v. (1936): Die Geochemie der Bariums. — Chemie der Erde, 10, 187—246, Jena.

Friedrich, O. M. (1953): Zur Erzlagerstättenkarte der Ostalpen. — Radex-Rundschau, 7/8, 371—407.

Friedrich, O. M. (1968): Die Vererzung der Ostalpen, gesehen als Glied des Gebirgsbaues. — Der Karinthin, 58, 6—17.

Gimm, W. (1947): Die magmatischen Lagerstätten des Thüringer Waldes. — Unveröffentl. Diss., Bergakademie Freiberg.

Gruškin, G. G. (1964): Einige Besonderheiten der Bildung von Fluorit-Lagerstätten (russ.). — Geologija rudn. mestorozd. 1, 15—32.

Jung, W. (1965): Zum subsalinaren Schollenbau im südöstlichen Harzvorland. Mit einigen Gedanken zur Äquidistanz von Schwächezonen. — Geologie, 14, 254—271.

Kopecký, L. (1971): Relationship between fenitization, alkaline magmatism, Barite-Fluorite Mineralization and deep-fault tectonica in the Bohemian Massif. — Upper Mantle Projekt Programme in Czechoslovakia 1962—1970, Geology Final Report, Praha *1971*, 73—95.

Krausse, H.-F. & A. *Pilger* (1969): Möglichkeiten der Rejuvenation von Pb-Zn-Erzlagerstätten im Saxonikum. — Meeting on Remobilization of Ores and Minerals. — Cagliari, August *1969*, 101—127.

Lange, P. & W. *Steiner* (1971): Eggische Strukturlinien im geologischen Bauplan der Deutschen Demokratischen Republik. — Geologie, *20*, H. 3, 213—235.

Leeder, O. (1966): Geochemie der Seltenen Erden in natürlichen Fluoriten und Kalziten. — Freib. Forsch.-H. C *206*.

Leeder, O. (1967): Die Einstufung von mitteleuropäischen Ganglagerstätten mit Hilfe des Gehaltes an Seltenen Erden. — Freib. Forsch.-H. C *209*, 99—117.

Mohs, A. (1931): Zur Frage der tertiären telemagmatischen Erze in Mitteleuropa. — Z. prakt. Geol., *39*, 13—15, Halle.

Oelsner, O. (1956): Zur Frage der Entstehung der saxonischen Lagerstätten, speziell auf den Randspalten des Thüringer Waldes. — Geologie, *5*, 685—694, Berlin.

Petrascheck, W. E. (1965): Typical features of metallogenetic provinces. — Econ. Geol., *60*, 1620—1634.

Petrascheck, W. E. (1968): Kontinentalverschiebung und Erzprovinzen. — Mineral. Deposita (Berl.) *3*, 56—65.

Sattran, V. & al. (1966): Problémy metalogeneze Ceského masivu. — Sborník geol. věd. ř. LG. *8*, 7—112.

Ščeglov, A. D. (1969): Endogenous deposits of the regions of autonomous activization (engl.). — Internat. Geol. Congr., Rep. of XXIII. Session Czechosl., Proc. Sect. 7, Endogen ore dep., 43—55.

Schneiderhöhn, H. (1941): Lehrbuch der Erzlagerstättenkunde. Bd. 1. — G. Fischer Verl., Jena.

Schneiderhöhn, H. (1949): Schwerspat- und pseudomorphe Quarzgänge in Westdeutschland. — N. Jb. Min., Mh., 191—202, Stuttgart.

Schneiderhöhn, H. (1953): Fortschritte in der Erkenntnis sekundär-hydrothermaler und regenerierter Lagerstätten. — N. Jb. Min., Mh., Stuttgart, 223—237.

Schröder, N. (1970): Die magmatogenen Mineralisationen des Thüringer Waldes und ihre Stellung im varistischen und saxonischen Mineralisationszyklus Mitteleuropas. — Freib. Forsch.-H. C *256*, Leipzig.

Schröder, N. (1971): Zur Gliederung, zeitlichen Einstufung und genetischen Deutung der magmatischen Mineralisationen Mitteleuropas — ein Diskussionsbeitrag. — Freib. Forsch.-H. C *288* (Top. Rep. of IAGOD, Vol. III), 121—137.

Smirnov, V. I. (1970): Geologie der Lagerstätten mineralischer Rohstoffe. — Übers. aus dem Russ.; Originalausgabe Moskau 1965). — VEB Deutscher Verl. f. Grundstoffindustrie, Leipzig. 563 S.

Staub, A. W. (1928): Beiträge zur Kenntnis der Schwerspat- und Flußspatlagerstätten des Thüringer Waldes und des Richelsdorfer Gebirges. — Z. deutsch. geol. Ges., *80*, 43—96, Berlin.

Staub, A. W. (1929): Die Flußspatlagerstätten des Thüringer Waldes. — Z. prakt. Geol., *37*, 49—55, 68—72, Halle.

Teuscher, E. O. & W. *Weinelt* (1972): Die Metallogenese im Raume Spessart-Fichtelgebirge-Oberpfälzer Wald-Bayerischer Wald. — Geologica Bavarica, *65*, 5—73, München.

Thienhaus, R. (1941): Die Schwerspatgänge des Richelsdorfer Gebirges. — Z. angew. Min., *3*, 21—52, Berlin.

Tuttle, O. F. & J. *Gittins* (1965): The *Carbonatites*. — Interscience Publishers, New York-London-Sydney. 591 S.

Werner, C.-D. (1958): Geochemie und Paragenese der saxonischen Schwerspat-Flußspat-Gänge im Schmalkaldener Revier. — Freib. Forsch.-H. C *47*, Berlin.

Werner, C.-D. (1966): Die Spatlagerstätten des Thüringer Waldes und ihre Stellung im Rahmen der saxonischen Metallprovinz. — Ber. deutsch. Ges. geol. Wiss., Reihe B, *11*, H. 1, 5—47.

Basic Geochemical Data of Zn, Pb and Cu and Hydrothermal Ore Genesis

K. H. *Wedepohl*[*]

Abstract

An efficient transport and a local accumulation of the ore metals is controlled by: 1. availability of water, 2. availability of metals and sulfur, 3. availability of compounds forming complexes with ore metals (chloride, bisulfide), 4. existence of channels and driving forces, 5. existence of steep geothermal gradients or solutions which cause ore precipitation. A concentration of metals at and above 1 ppm must be achieved in ore solutions (information from recent ore brines, fluid inclusions and solubility data of ore and gangue minerals). The availability of complexing chloride has been estimated from data on metamorphic and granitic rocks. Certain ore deposits can be correlated with magmatic, sedimentary or both sources by their lead isotope data.

Zusammenfassung

„*Geochemische Grunddaten von Zn, Pb und Cu und hydrothermale Erzentstehung*"

Transport und lokale Anreicherung von Erzmetallen werden durch folgende Faktoren kontrolliert: 1. das Vorhandensein von Wasser, 2. das Vorhandensein von Metallen und Schwefel, 3. das Vorhandensein von Verbindungen, die mit den Erzmetallen Komplexe eingehen (Chloride, Bisulfide), 4. das Vorhandensein von Kanälen für die Lösungen und von treibenden Kräften, 5. die Existenz von steilen geothermischen Gradienten oder Lösungen, die eine Erzausfällung bewirken. In Erzlösungen muß eine Metallkonzentration von etwa 1 ppm oder dar-

[*] Prof. Dr. K. H. *Wedepohl*, Geochemisches Institut der Universität, Goldschmidtstraße 1, D-34 Göttingen, FR Germany.

über erreicht werden (Ergebnisse von rezenten, metallreichen Solen, Flüssigkeits-einschlüssen und Löslichkeitsdaten von Erz- und Gangmineralen). Auf das Vor-handensein komplexierender Chloride wurde aus Daten von metamorphen und granitischen Gesteinen geschlossen. Bestimmte Erzlagerstätten können auf Grund ihrer Blei-Isotopen magmatischer oder/und sedimentärer Herkunft sein.

As in modern petrology concepts about the formation of ore deposits should rely on field observation, experiment and theory. Observations on geologic set-ting, on mineral composition and intergrowth as well as computations and measurements of mineral solubilities contribute to the knowledge of ore metal solutions. Recent sampling of ore brines in the Salton Sea area (California) and the Red Sea holes, progress in sophisticated analytical methods and experiments in sulfide systems have substantially influenced our views.

In this report I will consider the temperature range above early diagenesis of sediments and below the melting minimum of granitic rocks ($\sim 700^0$ C) as that of "hydrothermal solutions".

Several properties of a natural system are required for an effective transport and deposition of ore metals:

a) Availability of hot water or steam

b) Availability of metals and their sulfide anions

c) Availability of complexing compounds which change the very low solubility of metal sulfides drastically

d) Occurence of channels and driving forces for water circulation

e) Occurrence of steep gradients of temperature, water and/or rock compo-sition to dissolve and precipitate metals within short distances

a) *Availability of water* is often no problem in the middle and upper continen-tal earth's crust.

b) *Availability of metals (I) and sulfides (II)*

I. The abundant rock types and magmatic melts occurring in the continental crust contain Zn, Cu and Pb in concentrations mainly in the range of 10 to 100 ppm. The average abundance of these metals in granitic rocks, greywackes and shales (and their metamorphic equivalents) decreases in the sequence: Zn, Cu and Pb as visible in several frequency diagrams of Figure 1 plotted from data of a worldwide sampling compiled by *Wedepohl* (1972, 1973). One can observe that Zn, Cu and Pb occur almost everywhere in the continental crust in reasonable quantities to support solutions. The problem is to get these metals extracted into hydrothermal solutions. According to leaching experiments of andesite and shale by *Ellis* (1968) the proportion of extracted Cu and Pb (no data on Zn) increases with temperature and salinity of the solution and decreases with sulfur content of the rock. The almost complete extraction of Zn and Cu from Icelandic tholei-itic basalts by acid hydrothermal waters (pH 1.5) and the partial extraction of these metals by alkalic hot waters (pH 9.5) has been investigated by *Wedepohl* (1972; Fig. 30-G-1).

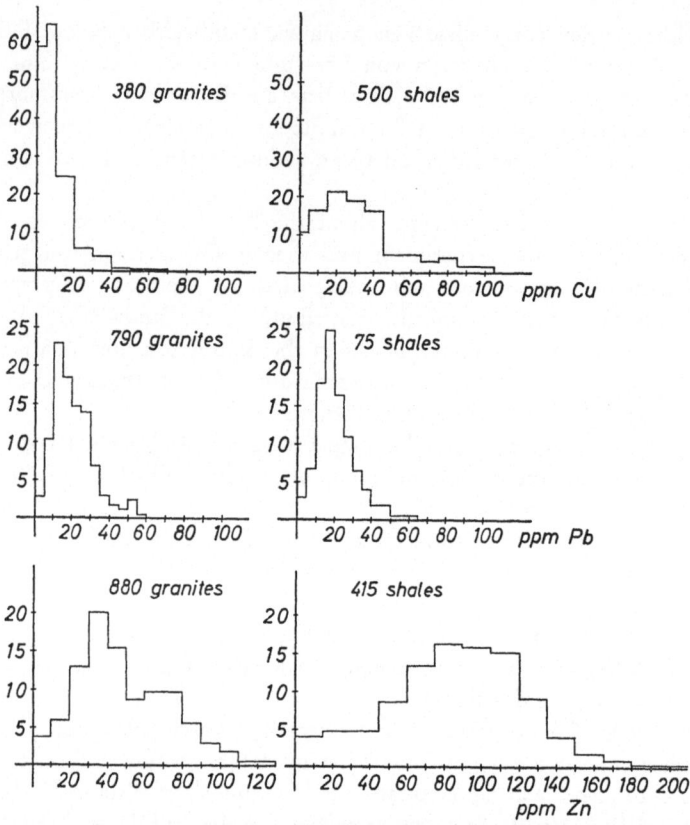

Fig. 1. Frequency distribution of Cu, Pb and Zn in granites and shales according to the compilation by *Wedepohl* (1972, 1973).

In magmatic rocks, zinc mainly occurs in the $Fe^{2+}Mg$ structural positions of biotite, amphibole and magnetite; copper is abundant as chalcopyrite and in the structure of rock forming $Fe^{2+}Mg$ minerals; lead has a strong tendency for the potassium position in feldspars and mica. In sedimentary rocks, Zn, Cu and Pb are incorporated in clay minerals; Zn and Cu partly occur in ferric iron oxides. Copper sulfides are more common in sediments than sphalerite and galena. Support of ore metals in hydrothermal solutions has to be expected from the consolidation of magmatic melts and from the extraction of wall rocks at fissures, voids etc. The different stability of the Zn, Cu and Pb host minerals against decomposition or dissolution in hydrothermal solutions partly controls the availability of these metals in ore depositing fluids. Chalcopyrite will be the least soluble of the mentioned host minerals of common rocks. All the three metals can be readily extracted from biotite, hematite etc. by reducing solutions. This fact can partly explain the predominance of sphalerite and galena over copper sulfides in numerous hydrothermal deposits.

II. Sulfide sulfur is abundant in granitic rocks in a concentration range of 100 ppm to 400 ppm S. Geosynclinal sediments contain on average 1000 ppm S *(Ronov & Yaroshevskiy,* 1967). The sulfur content of metamorphic rocks is expected to be closer to that of granitic than of argillaceous rocks. Reduced sulfur mainly occurs in the from of iron sulfides in common rocks. Because of the very low solubility of these sulfides hydrothermal waters need higher ionic strength, complexing compounds etc. to extract some sulfide from rocks and melts. Sulfide sulfur can be accumulated in a high temperature environment from a condensation of magmatic gas and at the low temperature level from bacterial sulfate reduction. Hydrothermal solutions with modest sulfide concentrations ($< 0,1$ m Σ $HS^- S^{2-}$) and low in chloride cannot extract Cu, Pb and Zn from rocks and melts because of the very low solubility of the respective sulfides.

According to *Krauskopf* (1961) the solubility of Cu_2S, PbS and ZnS in pure water increases by several orders of magnitude at an increase of temperature from 25^0 C to 300^0 C. But even at 300^0 C dissolved metals in equilibrium with sulfides are still as low as 10^{-7} to 10^{-9} mole per liter (Cu_2S: 10^{-9} mole/liter; ZnS 10^{-8} mole/liter; PbS 10^{-7} mole/liter). Water vapor of this temperature low in sulfide and chloride can dissolve metal concentrations of less than about 10 ppb Zn, Pb or Cu from rocks and melts. These concentrations are in the range of normal sea and river water; they are too low for ore metal accumulation in economic deposits.

Several estimates can demonstrate the inefficiency of solutions containing metals in the X to X0 ppb range to form deposits. In this case an ore body containing 10^6 to 10^7 tons Zn, Pb and/or Cu will need about 10^{15} tons of water as transporting agent. Twice the amount of water presently contained in all the rivers of the world has to pass through the fissures of a rock volume in the size of cubic kilometers ($n \cdot 10^9$ tons) within a reasonable time. The time is controlled by the tectonic activity which has to keep the fissures open.

Hot water and water vapor passing through rock fissures will dissolve several rock constituents as Si, Ba etc. from compounds which have usually a higher solubility than the minerals containing Zn, Pb and Cu. If one compares the abundance of common gangue and ore minerals of hydrothermal deposits and knows the solubility of quartz, barite etc. the range of ore metal concentrations can be estimated. The solubility of barite depends on temperature, sodium chloride and sulfate concentration (Figure 2a, according to *Templeton,* 1960 and *Strübel,* 1967), that of quartz on temperature and pressure (Figure 2b, according to *Kennedy,* 1950). The solubility of calcite has not been used because it is related to four parameters (temperature, P_{co2}, P_{total} and NaCl concentration) and decreases drastically from 100^0 towards 300^0 C *(Ellis,* 1959, 1963).

The solubility of quartz and barite in hydrothermal solutions is of special importance in the temperature range of 200^0 to 400^0 C and at a salinity of sea water and higher. These conditions have been recorded in numerous investigations of paleotemperature and fluid inclusions of vein type deposits. It can be

163

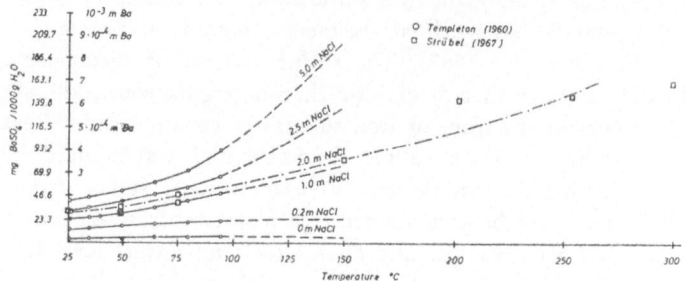

Fig. 2a: Solubility of barite as a function of temperature and NaCl concentration according to Templeton (1960) and *Strübel* (1967).

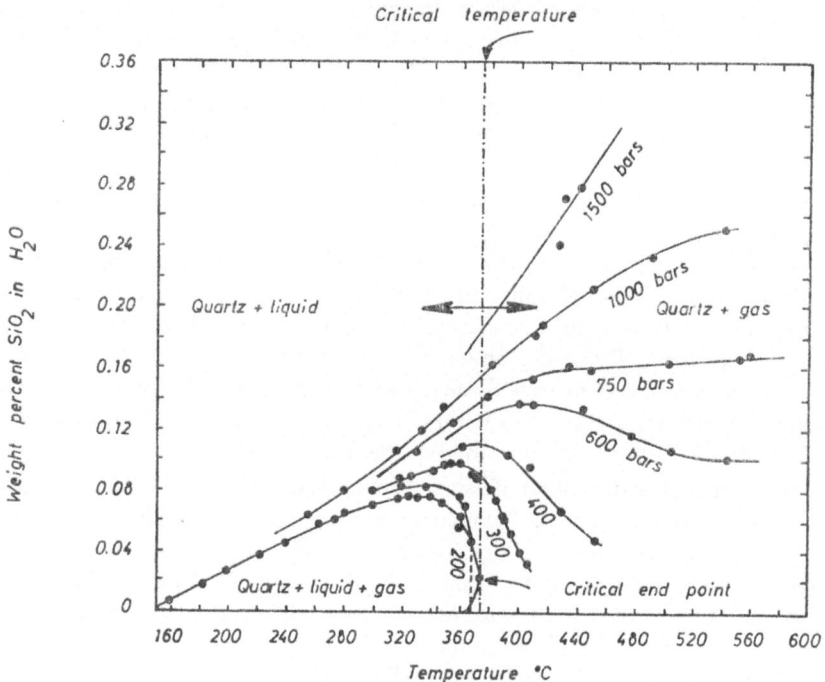

Fig. 2b: Solubility of quartz as a function of temperature according to *Kennedy* (1950). The influence of NaCl on the solubility can be neglected.

assumed that hydrothermal solutions are often saturated with respect to quartz and baryte because of the abundance of these minerals in common country rocks.

From Figures 2a + b it can be derived that at a temperature of 300⁰ C, a pressure according to that of 3 km depth, a sodium chloride concentration of sea water and a sulfate concentration of about 100 ppm: 1000 ppm quartz and 100 ppm baryte will be in solution. Only the concentration of dissolved BaSO₄ changes with salinity. An equal increase of the NaCl and the sulfate concentration has opposite effects which can almost compensate each other. If a ratio of quartz to ore minerals of 1000 and of baryte to metal sulfides of 100 (to 10) can be observed in hydrothermal deposits these are probably formed from solutions containing about 1 ppm of ore metals. These conditions and even lower gangue to ore mineral ratios (related to higher than 1 ppm metal concentrations in solution) are expected to have occurred abundantly in nature. A much lower metal concentration in the ppb range would be connected with a tremendous quartz mobilization. In our example of a relatively small ore deposit of 10^6 to 10^7 tons formed at 300^0 in 3 km depth a correlated mobilization of 10^{12} tons (300 km³!) of quartz is to be expected, an unreasonable mass transport.

Metal concentrations at and above 1 ppm have been oberserved in nature (Table 1) as for instance: 2—3 ppm Cu, 80 ppm Pb and about 500 ppm Zn in a brine of about 320^0 C at the Salton Sea (California) (*Helgeson*, 1968); 0.3 ppm Cu, 0.6 ppm Pb and 5 ppm Zn in a submarine brine at Atlantis II Deep in the Red Sea (*Emery, Hunt* & *Hays*, 1969); 10 ppm Pb in a highly saline brine in the Cheleken district (*Lebedev*, 1967) etc. up to 9000 ppm Cu and 1 % Zn in fluid inclusions of gangue minerals from North American deposits (*Czamanske, Roedder* & *Burns*, 1963).

In these solutions Zn is generally more abundant than Pb and Pb more than Cu. The sequence in abundance can be controlled by differences in solubility of the respective sufides (see Table 2) or by the different abundance of the metals in common rocks and magmatic melts (Figure 1: $Zn > Pb \geq Cu$).

c) *Availability of complexing compounds*

The very low solubility of sulfides in pure water and an increase of solubility in complexing experiments make it highly probable that complexing constituents occur in those hydrothermal solutions which form economic metal deposits. Appreciable metal concentrations can be transported as chlorides and/or hydrogen bisulfide complexes to form sulfide bodies (*Helgeson*, 1969; *Hemley* & al., 1967; *Barnes* & *Czamanske*, 1967). Results of their experiments on sulfide solubility are listed in Table 2 which prove that concentrations at and above 10 ppm Zn, Pb, Cu are attained under conditions which can have existed in nature. *Holland* (1972) reports that the quench of aqueous NaCl solutions, equilibrated with granitic melts at 800—850⁰ C, has a pH between 1.4 and 2.2 and *Hemley* & al. (1967) have measured in comparable systems (potassium feldspar-muscovite-quartz-saline solution) at 200⁰ to 300⁰ C a pH of about 4 to 5. The experiments by *Barnes* (reported in *Barnes* & *Czamanske*, 1967) and by *Romberger* & *Barnes* (1970) on bisulfide complexing have been performed in a weakly acid to alkalic pH range. For a significant metal transport of Zn, Pb and Cu as bisulfide com-

165

Table 1

Metals in hydrothermal waters

Water type, locality	Temperature [pH]	NaCl in g/l	Sulfide in ppm	Zn in ppm	Pb in ppm	Cu in ppm	Reference
Brine Salton Sea, Calif. (USA) Shell 2 Imp. Irr. Distr.	~320° C	208	30	500	80	3	Helgeson (1968)
Brine Red Sea Atlantis II Deep	56° C	249	n.d.	5	0.6	0.3	Emery, Hunt and Hays (1969)
Brine Caspian Sea (USSR) Cheleken	80° C	241	—	5.4	10.7	3.7	Lebedev (1967)
Hydrothermal water Mexico Cerro Prieto Mexicali	340° C	12—18	180—700	0.006	0.005	0.005	Mercado (1966, 1967), for reference: Ellis (1968)
Hydrothermal water Taupo, New Zealand Well 2 Broadlands	275° C	1.9	150	0.007	0.002	0.15	Ellis (1968)
Hydrothermal water Weirakei, New Zealand Well 48	250° C	2.4	12	0.009	0.002	0.06	Ellis (1968)
Hydrothermal water Auckland, New Zealand Well 1 Ngawha	230° C	1.5	2700	n.d.	0.01	0.02	Ellis (1968)
Formation waters Western Canada sedimentary basin	20—120° C [7.2]	91 [1—460]	n.d.	0.8 [0.03—28]	n.d.	0.12 [0.02—0.5]	Billings, Hitchon and Shaw (1969)

Table 2

Solubility of sulfides in complexing solutions in ppm of metal
(c: calculated; m: measured)

Metal (compound)	Complexing compound	100° C	200° C	300° C	500° C	Reference
Zn (ZnS)	3 mol. NaCl	0.29 c	8.2 c	205 c	n. d.	Helgeson (1969)
Zn (ZnS)	0.5 mol. KCl silicate buffer a)	n. d.	n. d.	65—650 m	2600—6500 m	Hemley et al. (1967)
Zn (ZnS)	2 mol. KCl silicate buffer a)	n. d.	n. d.	260—1800 m	3600—13 000 m	Hemley et al. (1967)
Zn (ZnS)	1 mol. HS- 0.5 mol. H_2S	n. d.	50 c	n. d.	n. d.	Barnes, Czamanske (1967)
Cu (Cu2S)	3 mol. NaCl	0.025 c	4.0 c	950 c	n. d.	Helgeson (1969)
Cu (CuFeS2)	3 mol. NaCl	0.034 c	0.64 c	12.7 c	n. d.	Helgeson (1969)
Cu (CuS)	1 mol. HS- 0.5 mol. H_2S	n. d.	180 c	n. d.	n. d.	Barnes, Czamanske (1967)
Cu (CuS)	0.66 mol. NaHS (+ H_2S)	25 m	240 m	n. d.	n. d.	Romberger, Barnes (1970)
Pb (PbS)	3 mol. NaCl	0.13	1.0	13	n. d.	Helgeson (1969)
Pb (PbS)	2 mol. KCl silicate buffer a)	n. d.	n. d.	620 m	16 600 m	Hemley et al. (1967)

a: silicate buffer: K-feldspar or pyrophyllite-muscovite-quartz and 1000 bar total pressure

plexes total sulfide concentrations almost as high as one molal are needed in neutral to weakly alkaline solutions.

The extraction of zinc by chloride complexing aqueous solutions from granitic melts has been proved experimentally by *Holland* (1972). The relation between the partition ratio of Zn (between vapor and melt) and the chloride concentration of the aqueous phase has been reproduced as Figure 3 from *Holland's* original diagram. The partition ratio increases from 2 to 60 as the chloride concentration increases from 1 to 5 moles NaCl per kilogram (equal to 5.8 to 29 weight percent). This indicates a strong partition into the aqueous phase. One molal chloride concentrations are common in liquid inclusions of quartz from hydrothermal deposits *(Roedder, 1967)*. Granite quartz from SW Germany and E France contains liquid inclusions with concentrations of 3 to more than 4 molal chloride according to *Goguel* (1963).

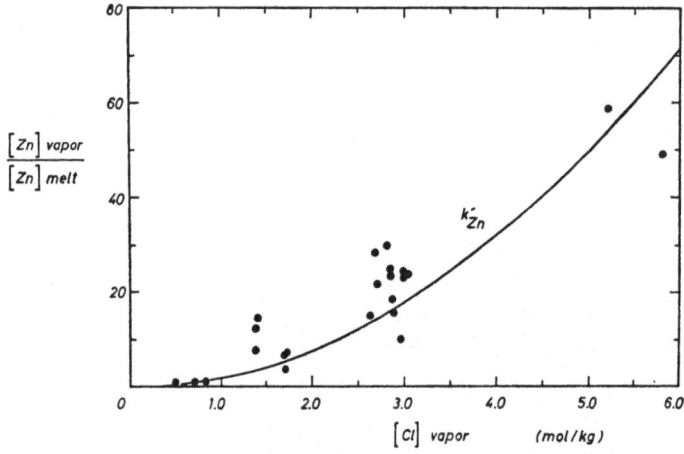

Fig. 3: The ratio of the concentration of zinc in the aqueous vapor to zinc in the granitic melt as a function of the chloride concentration in the aqueous phase according to *Holland* (1972). Plots of 22 experimental runs at 810—855° C temperature and 1.8—2.3 kb pressure. The curve represents a K'$_{Zn}$ of 9.5.

Because of the above mentioned significance of chloride complexing in hydrothermal metal transport ore formation with some probability is primarily controlled by the availability of chlorine. It can be expected that the maximum quantity of zinc that can be removed from a granite melt or a sedimentary country rock depends on their chloride content. In his model of zinc extraction from granites *Holland* (1972) used melts with 3500 ppm Cl. Average granites contain only 200 ppm Cl. But the largest proportion of their primary chlorine must have been lost during degassing and consolidation because chlorine favours the vapor phase during the partitioning between melt and gas (molal partition coefficient $K^{Cl} = m_{Cl}$ in melt/m_{Cl} in aqueous phase: 0.01—0.08 at 700—750° C and

168

2—8 kbars: *Kilinc* and *Burnham,* 1972). Therefore a knowledge of the initial chlorine concentration of granitic melts is important for our considerations.

Gneisses and mica schists of greenschist and amphibolite facies contain on average about the same chloride concentration as granites (as estimated from their mica, analyzed by *Haack,* 1969). *Johns* & *Huang* (1967) report a mean of 354 ppm Cl in 68 schists and 207 ppm Cl in 24 gneisses. During high grade metamorphism granitic melt can be formed in zones of migmatization. The fraction of partial melting of schists and gneisses will be controlled by their water content which is on average close to one percent. If partial melting occurs at 10 km depth (3000 bars pressure), about 8 % of water is soluble in a granitic melt, and a fraction of only 12 % melt will be formed in the metamorphic rock. In case of an extreme partitioning of chlorine into the melt its concentration will be 2000 ppm Cl. As a model melt *Holland* (1972) uses a granite containing 3550 ppm Cl and 1.8 to 5.4 % H_2O. Compared with the above computed chloride concentration in anatectic melts from gneisses and schitsts his model contains an appreciably higher chloride content. Deeper origin of partial melts, if saturated with water, is correlated with the formation of smaller fractions of melt and higher water and chloride concentrations in the melt. This consideration explains larger differences in primordial chloride concentrations of granitic melts which support a higher or lower chloride content in the vapor phase. According to the reported partition experiments only those granites, which are abnormally high in chloride, should be considered as potential sources of hydrothermal metal deposits.

As a consequence of his partition experiments *Holland* (1972) has computed the metal (Zn) loss from degassing granitic melts before and during consolidation. In case of his relatively high chloride concentration and depending on the primordial water content metal loss ranges from 20 to 100 percent of the original zinc content of the granitic magma. If a metal loss of more than 20 percent is to be expected, systematic differences between intrusive and volcanic rocks in Zn, Pb and Cu should be observed. Recent compilations of data on zinc and copper can be used for a comparison of granites and rhyolites (obsidians etc.) *(Wedepohl,* 1972, 1973). The average zinc concentration in 1106 granitic rocks and in 234 salic volcanic rocks (rhyolites, pitchstones and obsidians) is 50 ppm and 98 ppm Zn, respectively. The average copper concentration in more than thousand granitic rocks and 100 salic volcanics (rhyolites, obsidians, pitchstones) is close to 10 ppm Cu and 6 ppm Cu, respectively. Data on lead are similar for intrusives and extrusives. The zinc balance which should be used with some caution yet supports an assumption of a metal loss during consolidation of granitic melts. But most of this zinc probably will diffuse and not be concentrated in ore deposits.

Another problem is that of an evidence for special metal concentrations in those granites which are associated with ore deposits. *Putman* & *Burnham* (1963) did not observe any association between certain zinc and lead concentrations in intrusives and the occurrence of mineral deposits in granite areas of Arizona (USA).

169

But they found abnormally high copper in intrusives correlated with mineralization. *Parry & Nackowski* (1963) claim that abnormally low Zn-, Pb- and high Cu-concentrations in biotites from monzonitic intrusives in the Basin and Range stocks of Utah and Nevada are indicators of economic sphalerite, galena and copper sulfide deposition. The stocks containing copper above but lead and zinc below a certain level show characteristics of deuteric alteration. *Gavrilin & al.* (1967) describe correlations of albitization and sericitization of granitic rocks with hydrothermal mobilization of Pb und Zn.

Subsolidus mobilization of zinc during muscovite formation in granite has been observed by *Haack* (1969). The zinc concentration in biotite is positively correlated with the muscovite/biotite ratio in granites (biotite of biotite granites contains on average 290 ppm Zn; biotite of two mica granite with muscovite $>$ biotite: 520 ppm Zn).

The evidence presented in the above paragraph that lead zinc mineralization is indicated by the metal composition of related intrusives must be called meagre. This can have different reasons: Not at every intrusive from which a reasonable fraction of ore metals has been extracted during cooling, conditions favour a local precipitation to concentrate Zn and Pb into an ore body. Country rocks around an intrusive, penetrated by saline hydrothermal solutions, can contribute high fractions of a metal deposit. In certain mineralized areas magmatic plutons are even completely absent. At some of these places highly saline formation waters containing ore metals at the ppm level occur as in the Western Canada sedimentary basin. *Billings & al.* (1969) report concentrations of 20 ppm Zn and 0.5 ppm Cu in these waters as most probably being leached from shales.

Lead isotope data, as those compiled and partly contributed by *Doe* (1970), can locally solve the problem of the origin of lead in certain deposits. A direct magmatogenic origin must be concluded from the conformity between the isotopic composition of lead from a magmatic rock (or its feldspar) and that of galena from a related deposit. These ores are usually isotopically uniform. As examples, *Doe* (1970) mentions Butte (Montana, USA), Bingham Canyon (Utah, USA) and the Nelson batholith (British Columbia, Canada) etc. This group includes types in which at least a fraction of ore metal is derived from wallrock alteration of the outer zones of a pluton which are already consolidated. This stage can be characterized by sericitization mentioned above.

In lead ore being partly or completely supported by the lead of a country rock the isotopic composition of lead is usually more radiogenic and variable than that of the igneous body. The latter can have supplied complexing solutions, heat and a fraction of the metals. This case seems to be abundant.

For another type of ore formed at elevated temperature, any close association with magmatism can be excluded. It is usually characterized by highly radiogenic isotope ratios which in some cases can be traced back to rocks of the basement. The Mississippi-Valley type ore deposits belong to this association. Radiogenic lead is usually more easily to be extracted during wallrock alteration because of its occurrence in metamict minerals or in structural positions favourably to uranium and thorium instead of lead etc.

d) Channels and driving forces

The opening of deep faults, fissures and voids through which hydrothermal solutions can move, are often restricted to major tectonic events and belts. The relation between tectonism and magmatism has probably led to an overestimation of magmatic sources of ore metals in the past. Recent examples of closing of steam pipes in hydrothermal areas demonstrate the high rate of mineral deposition from solutions containing compounds with modest solubility. The closing of fractures by mineral deposition has to be compensated by the tectonic movement to get a deposit of reasonable size. In several vein type deposits tectonic movement during deposition is visible in brecciation of ore. The present author has estimated the time needed for the deposition of the three about 15 km thick Pb-Zn-Cu ore lenses of the Rammelsberg deposit (NW Germany) being a few thousand years only. This is a syngenetic sedimentary bed containing about $7 \cdot 10^6$ tons of metals most probably formed from submarine hydrothermal brines. The estimate of 0.3 cm per year as rate of deposition is based on the proportion of ore to silicate and the assumption of a reasonable rate of detrial accumulation in shallow seas (0.3 mm per year). The accumulation of the recent metalliferous Red Sea sediment ($>10^8$ tons ore) has been estimatd from ^{230}Th and ^{231}Pa growth as being at least 0.4 mm per year *(Ku,* 1969). The order of magnitude cm per year has been measured in studies of the movement of the present ocean floor. The time of deposition in single ore veins ($x \cdot 10^4$ to $x \cdot 10^5$ years) is shorter than the cooling time of a granitic intrusion of a reasonable size ($x \cdot 10^6$ years).

Outside areas of tectonic activity dissolution of country rocks can form channels through which hydrothermal solutions move. A chloride content of the solutions increases the solubility of $CaCO_3$ appreciably *(Ellis,* 1963). Mississippi-Valley type deposits often occur in limestones.

The most effective driving forces for the movement of hydrothermal solutions are gradients of temperature and pressure to allow convection. Local pressure gradients can even occur between pore space and fractures. Steep temperature gradients are to be expected within and around contacts of magmatic intrusions and in "heat domes" of regional metamorphism. Notwithstanding a remarkable mobilization of structural water from clay minerals and of NaCl containing formation waters, regional metamorphism is commonly not concentrating ore metals locally. Ore minerals formed during regional metamorphism are often randomly distributed. The zonal distribution of mineral associations (telescoping) in the thermal gradient of magmatic intrusions indicates that at least the driving force for the movement of hydrothermal solutions is of magmatic origin.

e) Ore precipitation

Ore precipitation needs oversaturation of solutions with respect to Zn, Pb, Cu sulfides. Complexing experiments mentioned in previous paragraphs, demonstrate the important influence of temperature and composition of a solution on the solubility of metal sulfides. Magmatic intrusions at shallow depth cause the steepest possible thermal gradients. A rapid change of the composition of a metal solution is achieved by mixing with waters lower in complexing anions. Magmatic vapors, brines, formation waters, sea water, geothermal waters and fresh water contain different concentrations of chloride as an important complexing anion. Mixing of waters of different origin seems to be an effective mechanism to precipitate different sulfides over short distances. Data on stable isotopes (D/H) indicate the rarity of truly magmatic solutions. Very often the $^{34}S/^{32}S$ composition of gangue minerals (baryte etc.), even of vein deposits with magmatic affinities, is close to that of contemporaneous sea water. This can be explained by the participation of formation waters at least in the process of ore and gangue mineral precipitation.

Another cause of precipitation is the change in p_H of ore solutions by reaction with country rocks having a reasonable buffering capacity as limestones and clays.

References

Barnes, H. L. & *Czamanske*, G. K. (1967): Solubilities and transport of ore minerals. — In: *Barnes* H. L. (editor), Geochemistry of Hydrothermal Ore Deposits. Holt, Rinehart & Winston, New York.

Billings, G. K., *Kesler*, S. E. & *Jackson*, S. A. (1969): Relation of zinc rich formation waters in the Western Canada sedimentary basin. — Chem. Geology 4, 211.

Billings, G. K., *Kesler*, S. E. & *Jackson*, S. A. (1969): Relation of zinc rich formation waters, northern Alberta, to the Pine Point ore deposit. — Econ. Geol. 64, 385.

Czamanske, G. K., *Roedder*, E. & *Burns*, F. C. (1963): Neutron activation analysis of fluid inclusions for copper, manganese and zinc. — Science 140, 401.

Doe, B. R. (1970): Lead Isotopes. — Springer, Berlin, Heidelberg, New York.

Ellis, A. J. (1963): The solubility of calcite in sodium chloride solutions at high temperatures. — Am. J. Sci. 261, 259.

Ellis, A. J. (1968): Natural hydrothermal systems and experimental hot-water rock interaction: reactions with NaCl solutions and trace metal extraction. — Geochim. Cosmochim. Acta 32, 1356.

Emery, K. O., *Hunt*, J. M. & *Hays*, E. E. (1969): Summary of hot brines and heavy metal deposits in the Red Sea. — In: *Degens*, E. T., *Ross*, D. A. (editors): Hot Brines and Recent Heavy Metal Deposits. Springer, Berlin, Heidelberg, New York.

Gavrilin, R. D., *Pevtsova*, L. A. & *Klassova*, N. S. (1967): Behavior of lead and zinc in the hydrothermal alteration of intrusive rocks. — Geokhimiya 8, 954.

Goguel, R. (1963): Die chemische Zusammensetzung der in den Mineralen einiger Granite und ihrer Pegmatite eingeschlossenen Gase und Flüssigkeiten. — Geochim. Cosmochim. Acta 27, 155.

Haack, U. K. (1969): Spurenelemente in Biotiten aus Graniten und Gneisen. — Contr. Mineral. Petrol. 22, 83.

Helgeson, H. H. (1968): Geologic and thermodynamic characteristics of the Salton Sea geothermal system. — Am. J. Sci. 266, 129.

Helgeson, H. C. (1969): Thermodynamics of hydrothermal systems at elevated temperatures and pressures. — Am. J. Sci. *267*, 729.

Hemley, J. J., *Meyer*, C., *Hodgson*, C. J. & *Thatcher*, A. B. (1967): Sulfide solubilities in alteration — controlled systems. — Science *158*, 1580.

Holland, H. D. (1972): Granites, solutions and base metal deposits. — Econ. Geology *67*, 281.

Johns, W. D. & *Huang*, W. H. (1967): Distribution of chlorine in terrestrial rocks. — Geochim. Cosmochim. Acta *31*, 35.

Kennedy, G. C. (1950): A portion of the system silica-water. — Econ. Geol. *45*, 629.

Kilinc, I. A. & *Burnham*, C. W. (1972): Partitioning of chloride between a silicate melt and coexisting aqueous phase from 2 to 8 kilobars. — Econ. Geol. *67*, 231.

Krauskopf, K. B. (1961): Übersicht über moderne Ansichten zur physikalischen Chemie erzbildender Lösungen. — Naturwissenschaften *48*, 441.

Ku, T. L. (1969): Uranium series isotopes in sediments from the Red Sea hot brine area. — In: *Degens*, E. T., *Ross*, D. A. (eds): Hot Brines and Recent Heavy Metal Deposits. Springer, Berlin, Heidelberg, New York.

Lebedev, L. M. (1967): Contemporary deposition of native lead from hot Cheleken thermal brines. — Dokl. Acad. Nauk S. S. S. R., Earth Sci. Sect. *174*, 173.

Parry, W. T. & *Nackowski*, M. P. (1963): Copper, lead and zinc in biotites from Basin and Range quartz monzonites. — Econ. Geol. *58*, 1126.

Putman, G. W. & *Burnham*, C. W. (1963): Trace elements in igneous rocks, Northwestern and Central Arizona. — Geochim. Cosmochim. Acta *27*, 53.

Roedder, E. (1967): Fluid inclusions as samples of ore fluids. — In: *Barnes*, H. L. (editor): Geochemistry of Hydrothermal Ore Deposits. — Holt, Rinehart and Winston, New York.

Romberger, S. B. & *Barnes*, H. L. (1970): Ore solution chemistry III. Solubility of CuS in sulfide solutions. — Econ. Geology *65*, 901.

Ronov, A. B. & *Yaroshevskiy*, A. A. (1967): Chemical structure of the earth's crust. — Geochemistry Internat. (Transl. of Geokhimiya 11, 1285, 1967).

Strübel, G. (1967): Zur Kenntnis und genetischen Bedeutung des Systems $BaSO_4$-NaCl-H_2O. —N. Jahrb. Mineral. Monatsh. *1967*, 223.

Templeton, C. C. (1960): Solubility of barium sulfate in sodium chloride solutions from 25^0 to 95^0 C. — J. Chem. and Eng. Data *5*, 514.

Wedepohl, K. H. (1972, 1973): Chapters 30 Zinc, 29 Copper and 82 Lead. — In: Handbook of Geochemistry Vol. 2, Springer, Berlin, Heidelberg, New York.

Die Herkunft der Erzmetalle

Walther E. *Petrascheck**

Zusammenfassung

Der Hauptgrund für die Annahme, daß die Erzmetalle eher aus der kontinentalen Kruste als aus dem Erdmantel stammen, ist die Tatsache, daß sich metallogenetische Provinzen in den wandernden Kontinenten vererben. Ein weiterer Grund ist das Fehlen größerer Erzlagerstätten in Verbindung mit ozeanischen Basalten oder Plateaubasalten. Sogar die sauren Differenziate auf Island werden lediglich von unbedeutenden Sulfidgängchen begleitet. Andererseits muß die während der geologischen Geschichte immer mehr anwachsende Tonnage an Buntmetallen in den Erzlagerstätten mit orogenetischen Ereignissen in Beziehung gebracht werden, mit denen basisches Mantelmagma in die sialische Kruste eintrat. Die flüchtigen Bestandteile dieses Mantelmagmas, besonders Schwefel und Chlor, laugten Elemente wie Zn, Pb, Sb, Sn, W und andere aus den Silikaten der sialischen Kruste heraus und konzentrierten sie. Dieser Vorgang setzte vor allem dann ein, wenn sialisches Grundgebirge der Geosynklinalen aufgeschmolzen wurde, und sich dieses palingene Magma der Kruste mit juvenilem Mantelmagma vermischte. Die Möglichkeit der Extraktion von Metallen aus Silikaten durch Schwefel und Chloride wird durch jüngst durchgeführte Experimente von *Kullerud, Hesp* & al. aufgezeigt. Es ist bemerkenswert, daß Geosynklinaltröge mit sialischem Grundgebirge keratophyrische Laven, die oft reich an polymetallischen Lagerstätten sind, enthalten, während Tröge mit ozeanischem Untergrund ophiolithische Vergesellschaftungen enthalten, die nur Kupfersulfide führen. Die metallogenetische Linie, die einen Kupfergürtel auf der ozeanischen Seite von einem Bleigürtel auf der kontinentalen Seite abtrennt, wie dies jüngst von *Petersen, Laznicka* und anderen für Nord- und Südamerika, und seit mehreren Jahren vom Autor für das alpin-mediterrane Orogensystem gefolgert wird, wird verständlich, wenn man das Kupfer von einer eugeosynklinalen ozeanischen Kruste und das Blei-Zink von der kontinentalen Kruste bezieht. Die regionalen Unter-

* Prof. Dr. W. E. *Petrascheck*, w. Mitglied d. Österreichischen Akademie der Wissenschaften, Montanistische Hochschule, A-8700 Leoben, Austria.

schiede der einzelnen metallogenetischen Provinzen können in den meisten Fällen durch „primäre" Unterschiede im geochemischen background innerhalb dieser sialischen Kruste und einer späteren Wechselwirkung mit dem Mantelmagma erklärt werden. Auf dem Ozeanboden können keine geochemischen Provinzen deutlich voneinander abgegrenzt werden.

Abstract

"The Origin of Ore Metals"

The main reason for the assumption of an origin of the ore metals from the continental crust rather than from the mantle is the heredity of metallogenic provinces in the wandering continents. Another reason is the absence of remarkable ore deposits in connection with oceanic basalts or with plateau basalts. Even the acid differentiates in Iceland are accompanied by insignificant sulphide veinlets only. On the other hand, the increasing tonnage of base metals in the ore deposits of the world generated during the geological history has to be correlated with orogenetic events together with which basic mantle magma was added to the sialic crust. The volatiles of this mantle magma, particularly the sulphur and the chlorides extracted and collected Zn, Pb, Sb, W and others from the silicates of the sialic crust. This happened particulary when the sialic basement of geosynclines became molten and this palingenic magma of the crust became mixed with the juvenile mantle magma. Recent experimental work done by *Kullerud, Hesp* and others demonstrates the possibility of extraction of metals from silicates by sulphur and chlorides. It is noteworthy that geosynclinal troughs with sialic basement contain keratophyric lavas often rich in polymetallic deposits whereas troughs with oceanic underground carry ophiolitic assemblages with coppersulfides only. The metallogenic line, dividing a copperbelt on the ocean side, and a lead belt on the continental side, as established recently by *Petersen, Laznicka* and others for North and South America, and since many years by the author for the Alpine-Mediterranean mountain system, can be understood owing to the origin of the copper from an eugeosynclinal oceanic crust and of lead-zinc from the continental crust. The main regional differences causing the different metallogenic provinces in most cases can be explained by "primary" differences of the geochemical background within this sialic crust and a later interaction with the mantle magma. Geochemical provinces on the ocean floor cannot be clearly distinguished.

Die Frage nach der Herkunft der Erzmetalle wird oft alternativ gestellt: Kruste oder Mantel? Ich möchte die Frage und die zugehörigen Begriffe etwas spezifizieren: es soll diskutiert werden, welche Metalle stammen vorwiegend aus der Kruste und welche vorwiegend aus dem Erdmantel; ferner sollen Erdmantel + Ozeanische Kruste gegenüber der Kontinentalen Kruste betrachtet werden; schließlich soll die Frage nach der Entstehung der kontinentalen (sialischen)

Kruste selbst nicht behandelt werden, sondern sie soll nur jene geologischen Zeiträume umfassen, in denen eine sialische Kruste mit ihrem differenzierten Stoffbe-
stand bereits existierte.

Die Frage hat schon vor mehr als 40 Jahren eine Antwort erfahren, als V. M.
Goldschmidt von chalkophilen und von lithophilen Elementen sprach. Sie wurde
später von verschiedenen Autoren behandelt — vom Standpunkt der Lagerstät-
tenforschung in der breitesten Weise von V. I. *Smirnow* (1968), der basaltophile
(= juvenile), grantiophile (= assimilierte) und filtrierte (= migrierte) Elemente
unterschied. Sie stellte sich neu, als die Beziehungen zwischen Kontinentaldrift
und Erzprovinzen und schließlich zwischen Plattentektonik und Erzprovinzen
studiert wurden.

Unabhängig voneinander und fast gleichzeitig kamen R. D. *Schuiling* (1967)
und W. E. *Petrascheck* (1968) zu der Auffassung, das Erzprovinzen mit wieder-
holtem Auftreten derselben Metalle, also mit metallogenetischer Vererbung
(„Metallprovinzen") nur erklärt werden können, wenn der Vorrat der betreffenden Metalle in den wandernden Kontinenten und nicht in dem darunterliegenden
Erdmantel beheimatet ist. *Schuiling* hat deutlich gezeigt, daß die geochemische
Anomalie der Zinngürtel beiderseits des Atlantik alt und kontinuierlich angelegt
war und nur durch die Drift zerrissen wurde. Dasselbe gilt für die praekambri-
schen Goldgürtel und für die Pegmatitprovinzen im Bereich der östlichen Gond-
wanakontinente *(Petracheck, 1972)*.

Dabei sind die Lagerstätten dieser Metallgürtel von unterschiedlichem Alter
und teils vor, teils nach Einsetzen der Drift entstanden. Die Zinn-Wolfram-
Lagerstätten Süd-Amerikas sind teils praekambrisch, teils paläozoisch, teils jurassisch, teils tertiär; jene von Afrika sind teils alt-praekambrisch, teils jung-prae-
kambrisch, teils jurassisch. In der nördlichen Fortsetzung des Wolframgürtels von
Ost-Brasilien-Nigeria dürften nicht nur die herzynischen W-Bezirke von Portugal
und Frankreich-Cornwall liegen, sondern vielleicht auch die große silurische
Scheelitlagerstätte von Mittersill in den Ostalpen. Doch gilt, was *Schuiling* sagt:
aus dem Bereich der geochemischen Anomalie kann erst ein „geologisches Ereig-
nis" eine Lagerstätte machen — und zwar zu verschiedenen Zeiten.

Zu den Elementen Sn, W, Au, Nb, Ta, Be, U, die sich als alt eingesessene,
durch Drift transportierte kontinentale Elemente dokumentieren, gehört aber auch
das wegen seiner Affinität zum Schwefel als chalkophil bezeichnete Blei und ein
Großteil des Zinks. Wiederholt habe ich darauf hingewiesen, daß der südliche
Stamm des alpin-mediterranen Orogens von Spanien — Nordafrika über die südlichen Alpen, die Dinariden bis in die Ketten des Taurus sehr reich an Pb-Zn-
Lagerstätten ist, während die nördlichen Ketten des Orogens die meisten Cu-
Lagerstätten enthalten. Neben vielen mesozoischen und tertiären Pb-Zn-Lagerstätten gibt es auch solche herzynischen Alters (z. B.: Sierra Morena, Sardinien,
z. T. Marokko). In diesen Schollen des süd-europäisch-nordafrikanischen Krusten-
bereiches sind also Blei und Zink beheimatet. Der Metallinhalt der Lagerstätten
zwischen Gibraltar und Euphrat liegt in der Größenordnung von 20 Millionen

Tonnen Pb + Zn. Auch der berühmte „Silbergürtel" von J. *Spurr* ist ein solch kontinentales Inventarium.

Führt uns somit die Erkenntnis der gewanderten Kontinente in Verbindung mit der metallogenetischen Vererbung zur Vorstellung eines frühzeitigen Vorhandenseins sehr vieler Erzmetalle in der sialischen Kruste, so bringen uns andere Lagerstättentypen zur Annahme ihrer Herkunft aus dem basischen Magma des Erdmantels. Das gilt unbestritten für die Cu, Ni, Pt und Cr-Lagerstätten, die an die großen basischen Instrusionen in den alten Schilden gebunden sind (Sudbury, Bushveld, Great Dyke, u. a. m.). Das Verhältnis der Schwefelisotopen beweist eindeutig rein magmatische Abkunft.

Das Problem eines Mantelursprunges vieler Erzmetalle stellt sich aber auch bei der Betrachtung der Metallogenese in den Orogenen. Viele Autoren, als erster F. *Blondel* (1936), später C. J. *Sullivan* (1948), J. A. *Bilibin* (1955) u. a. haben darauf hingewiesen, daß die Menge der chalkophilen Buntmetalle Cu, Pb, Zn, Sb zunehmend in den jüngeren Orogenen in Erscheinung tritt und sie haben dies auch quantitativ zu belegen sich bemüht.

Nach J. A. *Bilibin* sind 71 % des Kupferinhaltes der Lagerstätten der Erde nach dem Praecambrium gebildet, entsprechend der Weltlagerstättenkarte P. *Lafitte* und P. *Rouveyrol* (1964) sind 80 % des Cu, 78 % des Pb+Zn, 80 % des Pyrit, 85 % des Sb und 86 % des W in relativ jüngeren Lagerstätten, also seit der letzten jungpraekambrischen Orogenese konzentriert. Beim Zinn beträgt der „junge" Anteil rund 96 % (J. *Sarcia*, 1966).

Darüber hinaus glauben J. *Pereira* und C. J. *Dixon* (1965) auch einen Trend in der Zunahme des bauwürdigen Metallgehaltes der Buntmetallagerstätten in jüngerer Zeit feststellen zu können. Diese für die Erzkonzentration so ergiebige Zeitspanne von rund 1000 Millionen Jahren umfaßt weniger als ein Drittel der überschaubaren geologischen Geschichte.

Die Konzentration der wichtigen Erzmetalle zu Lagerstätten im Laufe der Erdentwicklung ist in den beiliegenden kumulativen Diagrammen dargestellt. Die Werte, die den jeweiligen Lagerstättenbildenden Metallinhalt, bestehend aus bisheriger Produktion + Reserven, in einzelnen geologischen Zeitabschnitten darstellen, sind aus der erwähnten Karte von *Lafitte* und *Rouveyrol* abgeleitet worden.* Aus diesen Diagrammen ist ersichtlich, daß Cr, Ni, Ti, Pt, Au, U zu den alten Lagerstättenmetallen gehören, während Cu, Pb, Zn, W, Sn, Sb und Hg vorwiegend und successive in nach-proterozoischer Zeit in Form von Lagerstätten in Erscheinung traten. (Allerdings können diese Zahlenwerte, die an und für sich seit 1964 wohl einer Korrektur bedürfen, durch Abtragung der praecambrischen magmenfernen Lagerstättenniveaus und andererseits durch sedimentäre Überdeckung tieferer jüngerer Niveaus verfälscht erscheinen — ein Einwand, den H.

---

* Die Umrechnung der graphischen Mengendarstellung der Karte zu einer Tabelle entnehme ich einer freundlichen Zusendung von Herrn Prof. A. *Maucher* (München); deren Transponierung in die kumulativen Kurven hat auf meine Bitte Herr Dr. P. *Walser* durchgeführt.

Pelissonnier (1971, p. 333) im Zusammenhang mit der Frage einer zeitlichen Evolution der Kupferlagerstätten machte. Dennoch ist das quantitative Prinzip dieser Kurven zutreffend, da ja gerade in den mesozoisch-tertiären Orogenen auch reichlich heißthermale Lagerstätten des Cu, Zn, Sn, W, u. a. freigelgt vorkommen). (Abb. 1—4)

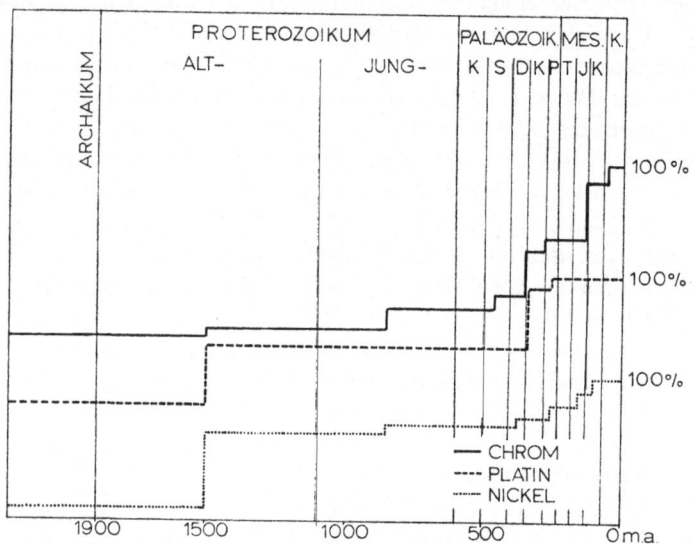

Abb. 1. Konzentration der alten Mantelmetalle Cr, Pt, Ni zu Lagerstätten im Laufe der Erdgeschichte.

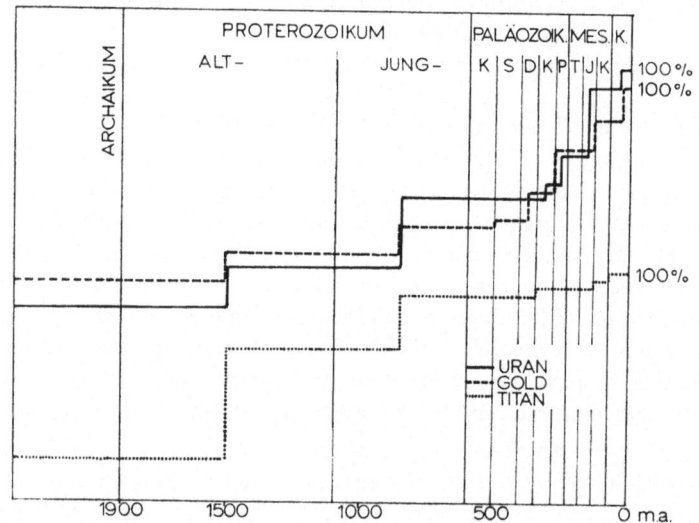

Abb. 2. Konzentration der alten Krustenmetalle U, Au, Ti zu Lagerstätten.

178

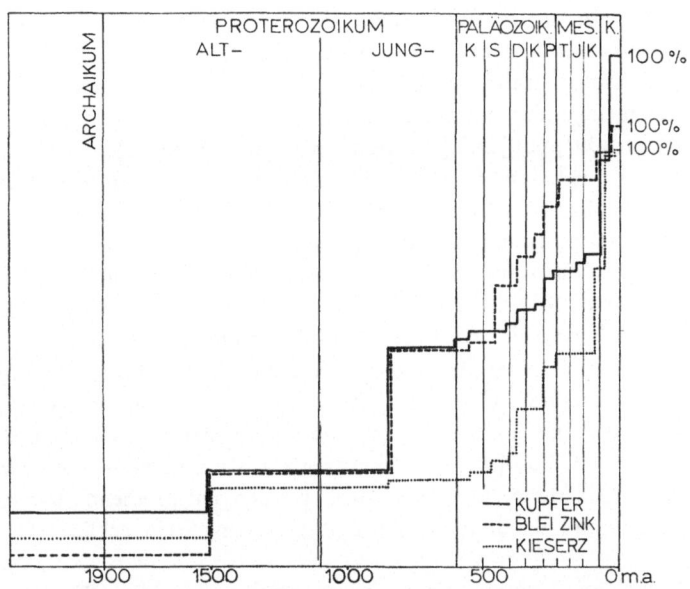

Abb. 3. Successive Konzentration der Buntmetalle Pb, Zn, Cu zu Lagerstätten.

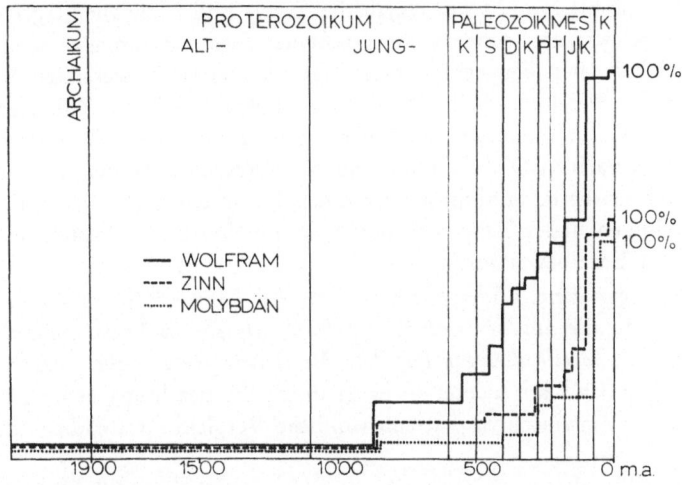

Abb. 4. Successive Konzentration der Krustenmetalle Sn, W, Mo zu Lagerstätten

Sullivan hat den lagerstättenbildenden Zuwachs der Buntmetalle mit der zunehmenden Beteiligung juvenilen basischen Magmas aus dem Mantel, bzw. der
ozeanischen Kruste erklärt. Diese Beteiligung ist aus den orogenen Vorgängen an
den unterschobenen Plattenrändern (consuming margins) verständlich.

179

Hier aber stellt sich ein Widerspruch ein, wenn man die aus weltweiter Erfahrung bekannte metallogenetische Sterilität der Basalte bedenkt — sowohl der ozeanischen Tholeitbasalte wie der kontinentalen Plateau- und Spaltenfüllungsbasalte. Zwar liegt der durchschnittliche Spurenmetallgehalt der Basalte bei Cu (87 ppm), Zn (39 ppm), Ni (200 ppm) wesentlich höher als bei Granit; bei Sn, W, Pb ist er niedriger. In dieses Durchschnittsbild passen die jüngst von S. *Jankovic* bekannt gemachten Werte von basischen Magmatiten in SE-Island mit Cu 20—300 ppm und Zn 10—200 ppm. Die saureren Differentiate (Rhyolit, Granophyr), die rund 10 % der magmatischen Gesteine SE-Islands einnehmen und die nach S. *Jankovic* (1972) als Differentiate des oberen Mantels aufzufassen sind, haben ein niedrigeres Cu-Maximum bei 20 ppm, aber ein ähnliches Zn-Maximum wie die Basalte; eine gründliche Prospektions-Campagne hat nur unbedeutende Gängchen mit Kupferkies sowie accessorischer Zinkblende und Bleiglanz im Bereich der Granophyrintrusionen aufgedeckt.

Das Wasser und die Metalle — überwiegend Mn und Fe — der warmen Salzsolen und Erzschlämme im Roten Meer werden entsprechend dem Deuterium-Gehalt exogen und durch Auslaugung von Nebengestein erklärt (D. E. White, 1968).

Der Mangel an Erzlagerstätten, die unmittelbar mit dem Basaltmagmatismus verknüpft sind, steht im Gegensatz zu zahlreichen vulkano-sedimentären Erzlagern, die mit dem basischen Geosynklinalvulkanismus in Zusammenhang stehen. Das sind bekanntlich nicht nur oxydische Eisen- und Manganerzlager mit ausgeschiedener Kieselsäure, sondern auch polymetallische Sulfidlager aller Art. Die Bereitwilligkeit, mit der heute keineswegs nur mehr Schwefelkies mit Kupfer, sondern auch Bleiglanz, helle Zinkblende, Fluorit, Stibnit, Zinnober, Scheelit, Molybdenit, Gold, Magnesit und Siderit, teils zusammen, teils monomineralisch auf einen submarinen Diabasvulkanismus zurückgeführt werden, hat bei manchen Lagerstättenforschern im alpin-mediterranen Raum ein großes Ausmaß angenommen. In einigen dieser Fälle war die vulkanosedimentäre Entstehung durch die Struktur der Erze beweisbar.

Der oben genannte Widerspruch scheint mir dann gelöst, wenn man die besprochenen Metalle aus der sialischen Kruste, die sie zu Erzen konzentrierenden flüchtigen Bestandteile dagegen aus dem Mantelmagma bezieht. Diesen Gedanken habe ich schon 1968 angedeutet — er ist durch die Beachtung neuerer experimenteller Arbeiten sowie durch plattentektonische Vergleiche wahrscheinlicher geworden.

Der Metallgehalt der sialischen Gesteine ist zumeist an silikatische Mineralien gebunden. Nach K. *Wedepohl* ist weit mehr Zink in den Mg-Fe-Silikaten der Gesteine als in den Lagerstätten enthalten. Olivin hat einige 1000 ppm Zink, Pyroxen, Amphibol und Biotit und einige 100 ppm. Zn. In Biotit ist Cu bis zu 1000 ppm enthalten, in Kalifeldspat Pb bis zu 10 ppm. Ebenso ist Sn im Biotit bis zu 300 ppm angereichert. Die basaltischen Magmen enthalten reichliches Cu. In submarinen ozeanischen Basalten wurden 700—900 ppm Schwefel bestimmt. (J. G. *Moore* und B. P. *Fabbi*, 1971). Vor allem aber exhalieren sie in Gasform neben Wasserdampf H_2S, CO_2, H_2 und Halogenide.

Wenn nun an den unterschiebenden Plattenrändern bei der Bildung von Kettengebirgen des Cordilleren-Typs Teile der kontinentalen Kruste zusammen mit der ozeanischen Kruste in die Tiefe gezogen und aufgeschmolzen werden, entstehen hybride Mischmagmen; ebenso entstehen solche, wenn in intrakontinentalen Geosynklinalen (wie etwa der herzynischen Geosynklinale Europas oder der alpidischen der Ostalpen und Karpathen), deren sialischer Boden in den palingenen Schmelzbereich eingetaucht wird. Dann können die in der kontinentalen Kruste — und natürlich auch in den geosynklinalen Sedimenten — diffus verteilten Metalle von den flüchtigen Bestandteilen des Mantelmagmas extrahiert und zu Lagerstätten konzentriert oder zumindest praekonzentriert werden.

Die Metallurgen verneinen zwar die technische Möglichkeit einer Gewinnung von Buntmetallen aus silikatischer Schmelze durch Schwefel. G. *Kullerud* und H. S. *Yoder* haben jedoch 1963 Fe und Ni aus Olivin mit Schwefel bei 800^0 C und 2000 bar zu Magnetkies und Pentlandit umgesetzt. Später konnten in einer Anzahl von Experimenten Fe, Ni, Co, Pb, Zn und Hg in Form von Sulfiden aus Silikaten und Carbonaten extrahiert werden (G. *Kullerud* & G. H. *Moh*, 1972). Sn wurde kürzlich von W. R. *Hesp* und D. *Rigby* durch eine NH_4 Cl-Lösung bei 500^0 C aus einem Biotit eines SW-australischen Granits mit 60 ppm Sn herausgelöst. Es ist also möglich, Erzmetalle aus Silikatmineralien und Silikatschmelzen durch flüchtige Bestandteile eines Magmas zu extrahieren. Immerhin enthält 1 km^3 sialische Kruste mehrere Zehntausend Tonnen der Buntmetalle.

Nach einer solchen Vorkonzentration der Metalle in der Kruste durch den Sammler-Effekt des Schwefels und anderer flüchtiger Bestandteile des Mantels treten jene „geologischen Ereignisse" ein, die *Schuiling* für die eigentliche Lagerstättenbildung voraussetzt. Diese Ereignisse können die ganze Variationsbreite der lagerstättenbildenden Vorgänge umfassen. Die dabei auftretenden Metalle kommen teils aus dem juvenilen Mantelmagma, darunter besonders das Kupfer, zum größeren Teil aber aus dem palingenen Magma der sialischen Kruste.

Damit deutet sich auch eine Erklärung für den erwähnten erzbringenden Geosynklinalvulkanismus an: die vulkano-sedimentären polymetallischen Lagerstätten finden sich vor allem an die spilitisch-keratophyrische Gesteinsgruppe geknüpft, während die rein ophiolithische höchstens Kupferlagerstätten brachte. Nun enthält nach dem Modell der Plattentektonik der kontinentseitige Trog mit sialischem Boden die Keratophyre, der ozeanseitige Trog mit basaltischem Boden die diabasischen Pillowlavas und eventuell hochgeschleppte Gabbroide und peridoditische Gesteine. Somit ist auch die reiche Erzführung des initialen Vulkanismus mancher Geosynklinalen — z. B. der variscischen Deutschlands — trotz der primären Sterilität des juvenilen basaltischen Magmas eigentlich von sialischer Herkunft.

Aus der ozeanischen Kruste dagegen, vor allem wohl aus ihrer differenzierten gabbroiden tiefen Lage 3, stammen das meiste Kupfer, Mangan und Eisen. *Pelis-*

sonier hat in seiner hervorragenden quantifizierten Übersicht über die Kupferlagerstätten festgestellt, daß 15 % der Welt-Tonnage von Kupfer direkt mit basischen Gesteinen verknüpft sind, aber 40—45 % — daruner eben die „Porphyries" — indirekt vom Mantel abzuleiten sind.

In dieses Bild der Herkunft der Buntmetalle fügt sich auch die neuerdings vielfach hervorgehobene Grenzlinie zwischen Kupferzone und Bleizone in den jüngeren Orogenen, besonders der beiden Amerikas ein (U. *Petersen*, 1964, P. *Laznicka* & H. D. B. *Wilson*, 1972). Die ozeanseitige Kupferzone wird vom Mantel, die kontinentseitige Bleizone von der Kruste abgeleitet. Ich selbst hatte eine solche Zonengliederung für das alpin-mediterrane Orogen seit 1955 betont und hatte diese Differenzierung mit einem primären Unterschied des Stoffbestandes der tieferen Unterkruste zu erklären versucht. Ich möchte dies heute umdeuten, indem die nördliche Kupferzone (mit großen Lagerstätten von Fe und Mg und accessorischen Begleitmineralien von Ni und Cr) mit der nördlich des alten afrikanischen Kontinents gelgenen penninischen Eugeosynklinale in Zusammenhang steht, während die angrenzenden Partien dieses afrikanischen Kontinentes eine hohe geochemische Pb-Zn-Anomalie aufwiesen und darum zu verschiedenen Zeiten die Geburtsstätte von Pb-Zn-Lagerstätten waren. Die spätere alpine Überschiebungstektonik hat diese räumlichen Beziehungen komplizierter gemacht.

Erze aus dem Erdmantel selbst haben wir nur dort, wo dieser durch Tektonik tief aufgepflügt wurde. So sind die alpinotypen Chromerzlagerstätten der von großen Deckenbewegungen erfaßten ophiolithischen Subduktionszonen der Balkanhalbinsel und Vorderasiens zu deuten. Ruhiger Aufstieg von Mantelmaterie in ozeanischen Rücken oder in Rift-Valleys bringt neben H_2S und CO_2 — Exhalationen vorwiegend nur Wärmestrom, der zur Metallauflösung aus Sedimenten und Ausfällung durch „sekundäre Hydrothermen" führt.

Die regionale Correlation der metallogenetischen Provinzen der Erde im Hinblick auf Plattentektonik und Drift zeigt, daß die großen Provinzen durch das Zusammenwirken von Kruste und Mantel gebildet wurden. Die stofflichen Unterschiede liegen primär vorwiegend in der kontinentalen Kruste begründet — die ozeanische Kruste scheint, wie E. *Ingerson* in seinem Vortrag bei diesem Symposium mit aller Vorsicht gesagt hat, gleichmäßig zusammengesetzt zu sein. Auch aus den Manganknollen ließen sich keine unterschiedlichen primären Provinzen abgrenzen — das Verhältnis Fe/Mn und das der Spurenmetalle ist von sehr vielen Faktoren bestimmt (D. S. *Cronan*, 1972).

Literaturhinweise

Cronan, D. S. (1972): Regional geochemistry of the ferromanganese nodules in the world ocean. — In: Ferromanganese deposits on the ocean floor, ed. D. R. Horn, Lamont Observatory New York.
Heirtzler, J. R. (Editor) (1972): Understanding of the Mid Atlantic Ridge. — Nat. Ac. Sc. Ocean Science Comm., Washington.
Hesp, W. R. & *Rigby*, D. (1972): The Transport of tin in acid igneous rocks. — Pacific Geology 4, p. 135—152, Tokyo.

Jankovic, S. (1972): The origin of base metal mineralization on the Mid Atlantic Ridge. — C. Rend. 24. Int. Geol. Congr. Sect. *4*, p. 326—335, Montreal.

Kullerud, G. & G. H. *Moh* (1972): Das Problem Erz-Nebengestein, erläutert an ausgewählten Beispielen im Experiment. — Min. Dep. *7*, p. 271—279, Berlin.

Laznicka, P. & *Wilson*, H. D. B. (1972): The significance of a Copper-Lead Line in Metallogeny, C. Rend. 24. Int. Geol. Congr. Sect. *4*, p. 25—36, Montreal.

Moore, J. G. & *Fabbi*, B. P. (1971): An estimate of the juvenile sulfur content of basalt. — Cont. Min.-Petrol. *33*, p. 118—127.

Pelissonier, H. (1972): Les dimensions des gisements de cuivre du monde. — Mem. du B.R.G.M. *57*, p. 1—405, Paris.

Petersen, U. (1972): Geochemical and tectonic implications of South American Metallogenic provinces. — Ann. New York Ac. Sc. *196/1*, p. 1—38.

Petrascheck, W. E. (1963): Die alpin-mediterrane Metallogenese. — Geolog. Rundsch. *53*, p. 376—389.

Petrascheck, W. E. (1968): Kontinentalverschiebung und Erzprovinzen. — Min. Dep. *3*, p. 56—65.

Petrascheck, W. E. (1973): Some aspects of the relations between Continental Drift and Metallogenic Provinces. — In: Implications of Continental Drift to the Earth Sciences, ed. by D. H. *Tarling* and S. K. *Runcorn*, Vol. 1, 567—572.

Schuiling, R. D. (1967): Tin belts on the continents around the Atlantic Ocean. — Econ. Geol. *62*, p. 540—550.

Smirnov, V. I. (1968): The sources of ore forming materials. — Econ. Geol. *65*, p. 380—389.

White, D. E. (1968): Environments of Generation of some base metal ore deposits. — Econ. Geol. *63*, p. 301—335.